NEUROLOGY - LABORATORY AND CLINICAL RESEARCH DEVELOPMENTS

MOLECULAR ASPECTS OF THE PSYCHOSOMATIC-METABOLIC AXIS AND STRESS

NEUROLOGY - LABORATORY AND CLINICAL RESEARCH DEVELOPMENTS

Additional books in this series can be found on Nova's website
under the Series tab.

Additional e-books in this series can be found on Nova's website
under the e-book tab.

NEUROLOGY - LABORATORY AND CLINICAL RESEARCH DEVELOPMENTS

MOLECULAR ASPECTS OF THE PSYCHOSOMATIC-METABOLIC AXIS AND STRESS

ALFRED BENNUN, PH.D.

NOVA BIOMEDICAL

New York

Library of Congress Cataloging-in-Publication Data

ISBN: 978-1-63463-912-5
Library of Congress Control Number: 2014960319

Published by Nova Science Publishers, Inc. † New York

Contents

Preface

It is desirable to ask questions that could lead to further developments in models describing the function of the human brain noradrenergic systems on the control by the brain of the associated body metabolic network. This leads to analyze development of brain through its metabolic self-function on the structuring of psychosomatic connectivity.

Hence, the initial question: *Could be that along evolution fight-or-flight lead to consolidate new experiences, required for survival and brain self-gratification, into memory mechanisms*?

This suggested that a model for memory and intelligence could be deducted from the brain noradrenergic system acting on the hypothalamic-pituitary-adrenal (HTPA) axis for control of the adrenals gland secretion of adrenaline in blood.

The latter relationship describes how the psychosomatic system could become integrated. The sensory neurons are capable to activate the locus coeruleus (LC). The LC-long axons by reaching almost all brain areas allow multiple sensory inputs to became consolidate into mental sceneries.

LC-neurons by simultaneously releasing noradrenaline (NA) activate postsynaptic adenylyl cyclases (AC) of cortex, hypothalamus, striatum and other areas. Activatory responsiveness to NA requires Gs protein coupling with AC. This one is dependent of free Mg^{2+} in excess of substrate Mg-ATP. Free Ca^{2+} activates Gi protein for the NA-receptor dependent inhibition of basal AC. Opposite responses characterize glutaminergic neurons, in which Mg^{2+} inhibits the NMDA receptors, whereas the entrance of Ca^{2+} become activatory.

Neurons and astrocytes are integrated cell systems modulated by ionic-metabolic parameters, dependent of the Mg^{2+} and O_2 supplied by the capillary.

The histidine fingers of Hb could chelate Mg^{2+} with the partial loss of the hydration shell. The oxy to deoxyHb conformational change releases a partially hydrated Mg^{2+} in a *nascent* more reactive state of greater manifest effective charge.

cAMP by activation of the voltage-gated ion channel could initiate the action potential. The less-hydrated state of the Mg^{2+} ion could capture H_2O from the hydration shells of Na^+ and K^+. This H_2O sequestration sequence decreases the ion-water diameter to match in term the size of its corresponding gate at the neuronal membrane. Activation of the hypothalamic NA-AC at the hypothalamic-pituitary-adrenal (HTPA) axis, leads to the adrenal gland release of cortisol and adrenaline into blood. Several hormones, adrenaline, ACTH, glucagon, etc., can favor the activatory Mg^{2+}-dependent adrenergic response of AC of several body tissues, shifting the relative rates of several metabolic pathways. Adrenaline stimulates the use of the stores of lipids and glycogen. Cortisol controls the amino acid release from organismal proteins to support gluconeogenesis. This configures an *in common* Mg^{2+} link allowing the noradrenergic system of the brain to self-control through the HTPA axis of the metabolic activity of its supporting tissue network.

Cortisol crosses the blood-brain barrier (BBB), but adrenaline cannot and the HTPA axis became a homeostatically incomplete axis, because could not completely turn-off the fight-or-flight response.

cAMP response element-binding (CREB) dependent DNA transcription allows coding neuronal circuits for consolidated long-term memory. The psyche/brain lack of an adrenaline negative feedback could acquire an unconscious level, mainly explored by psychoanalytic technics. Positron emission tomography has revealed the metabolic impact of anxiety and stress on specific locus/regions of the brain. This indicates the need for integrative studies to configure the molecular relationship between the mind and its dysfunctions.

Chapter 1

Methods

Mg^{2+}-Treated Particles of Adenylyl Cyclase

Male Sprague-Dawley (300g) were stunned and decapitated at 4°C. The cerebral of each brain was sliced and collected in 3ml of a 2mM-Tricine-Tris-1mM-$MgSO_4$ buffer, pH=7.4. The tissue was homogenized at low speed for 2min, and then centrifuged in a cold-room (4°C) at 1700 RPM for 20min. The pellet was washed twice and finally suspended in 3ml of the same homogenizing medium and stored at -70°C. After thawing for 1h at room temperature, portions of the particulate preparations were incubated under the conditions described under the assay of adenylyl cyclase activity [1].

Ca^{2+}-Treated Particle Cyclase Preparation

Male Sprague-Dawley rats weighing 200g were stunned and decapitated at 4°C. Cerebral cortex from several brains were sliced and collected in a volume (4ml/cortex) of ice-cold 10mM-Tricine/Tris buffer, pH=7.4, containing 2mM-sucrose. Then, homogenized using a motor-driven glass at 400 RPM for 2min and centrifuged at 2400 RPM for 20min in a 5°C cold-room. The supernatant fraction was discarded, the pellet was washed by resuspension with a vortex mixer in an equal volume of the same homogenizing medium, and recentriguged at 2400 RPM for 20min. A second washing was done by same procedure.

The pellet was suspended of 10mM-Tricine/Tris buffer, pH=7.4, containing 0.5mM-$CaCl_2$. The Ca^{2+}-treated particle was used either fresh indicated in the legend, or after storage at -70°C. Thawing for 30min at room temperature (24°C). Before its experimental use, the Ca^{2+}-treated adenylyl cyclase (AC) particulate preparation was washed by same centrifugation procedure and resuspended in the same medium at the desired concentration [2].

EDTA-Purified Membrane Preparations from Cortex, Hypothalamus and Striatum Brain Tissue

Female Sprague-Dawley rats weighing 210-300g were ovariectomized to decrease interference by estrus cycles [2] and decapitated at 4°C two weeks after surgery. The brains were dissected in the cold (4°C) to isolate and separately collect the hypothalamus, cortex and striatum from groups of 10 to 15 animals. These tissues were selected in expectation of finding differential kinetic properties, which could be related to specific areas of rat´s brain.

The total wet weight of each pool of tissues was colected in 20 ml of ice-cold 2 mM Tris-HCI containing 10 mM NaCl, 10 mM KCl and 5μM EDTA, buffer at pH=7.4. These were rinsed twice with 15 ml/g wet weight of tissue and homogenized using a motor-driven (Inframo, Wayne, New Jersey) Teflon homogenizer at 400 rpm for 2 min. Each homogenate was centrifuged at 4°C and 6,000 g_{av} in a RCA Sorvall, head SS-34 for 20 min. The resulting supernatants were discarded and the pellet washed, resuspended and centrifuged again once.

Aliquots stored at -70°C. Protein concentration for each membrane preparation was determined by the method of Lowry [3], using bovine serum albumin as the standard. Particle suspensions and reaction mixtures were tested with a calcium ion electrode for Ca^{2+} contamination with negative results [4]. These EDTA-treated adenylyl cyclase particulate preparations were stored at -70°C and used after thawing for 30min at 24°C [2]. The cyclic adenosine monophosphate (cAMP) formed was measured by a modification of the Gilman method [5].

Assay of Adenylyl Cyclase

Adenylyl cyclase (AC), systematic name: ATP diphosphate-lyase (cyclizing, 3´, 5´-cyclic-AMP-forming), EC 4.6.1.1. Defrosted preparations from the above mentioned differents treated membranes (from 1.7 to 6 μg of protein per assay) were re-suspended in a standard reaction mixture, containing 20 mM Tris-HCl buffer (pH 7.4), 5 mM theophylline, $MgCl_2$, plus ATP disodium salt, extracted from equine muscle, Sigma, a GTP content of about 2% and traces of other nucleotides.

This ATP containing GTP was used when the neurotransmitter noradrenaline-dependent activation was investigated. The ATP disodium salt from P-L Biochemicals, Inc (Milwaukee, Wisconsin) was used to obtain an assay of basal AC activity independent of the neurotransmitter activatory effect.

After incubating in a shaking bath at 37°C for 15 min the reaction was terminated by heating in a boiling water bath for 5 min. Time-dependence studies using the membrane preparations, showed linear responses for no less than 30 min.

Isolation of Cells from Fat Tissue

Male Sprague-Dawley [6] rats (about 2 months old and weighing 200-250 g) were stunned and decapitated and the epididymal adipose tissue was rapidly excised.

Isalated fat-cells were obtained as described by Rodbell [7]. The freshly excised weight, and suspended in 10ml of KRP buffer, pH=7.4 containing 2.5% free fatty acid-depleted albumin (per g wet tissue wt) and incubated at 37°C for 90 min with 10mg/g of wet tissue weight of crude collagenase (EC 3.4.24.3) (Clostridium hystolyticum, Worthington Biochemical Corp, Freehold, NJ, USA, type CLS). The dispersed cells were collected after centrifugation at 4°C and $400g_{AV}$ for 3 min and then washed twice by resuspension in KRP buffer (pH=7.4) without added albumin, followed by centrifugation under the previously described conditions [8].

Particulate Preparation of Adenylyl Cyclase from Adipose Tissue

Isolated fat-cells were resuspended with 10 ml/g wet. tissue weight. of an ice-cold hypotonic solution of 2 mM-Tris/HCl (pH=7.4), 10 mM-KCl, 10 mM-NaCl and 5 μM- EDTA and shaken 10 min at 4°C, and 1300 g_{av} centrifuged for 3 min. The floating fat cakes were removed, resuspended in the same buffer, shaken as previously described and centrifuged twice more. The pellets obtained from all these 3 centrifugations, were resuspended in 7 ml buffer/g w.t., and cooled to 4°C. Homogenization was performed with a Potter-Elvehjem homogenizer (Teflon pestle, 10 up-and-down strokes, clearance about 0.02 mm, 200 rev./min). The homogenates were centrifuged at 4°C and 30,000 g_{av} for 3 min. The pellets resuspended in 1 ml of the same buffer/g w.t. and homogenized, 0.5 ml portions in separate vials were then stored at -70°C. Stored preparations were stable for no less than 6 months. Adenylyl cyclase (AC) preparations were thawed slowly at 4°C before their immediate use.

The thawed enzyme suspension can he maintained at 4°C for 60min without any detectable loss of enzyme activity. The reaction was always initiated by addition of the enzyme to the reaction mixture containing all other reagents. MgATP prevents the inactivation of the enzyme by l-adrenaline and the reaction rate maintains linearity over longer times than 20min. The enzymatic rate is proportional to the enzyme concentrations even up to 5-fold above the used one.

The thawed unused portions were always discarded. The protein contents of these AC fat membranes were assayed by the method of Lowry *et. al* 1951 with bovine serum albumin as the standard [6].

Assay of the Particulate Fat Adenylyl Cyclase from Fat-Cell

Portions of the adenylyl cyclase (AC) preparations were incubated at 37°C for up to 20min in a standard reaction mixture containing 40 mM-Tris/HCl buffer (PH 7.4) 5mM-creatine phosphate, 0.25 mg/ml creatine kinase, 5mM-theophylline. ATP and $MgCl_2$ were always previously neutralized to the same pH 7.4. Incubation was terminated by heating in a boiling water bath for 4 min. Zero time blanks were obtained for all assays. Upon preincubation with l-adrenaline at 37°C prior to the addition of substrate, the AC activity of the particulate preparation decayed rapidly, paralleling similar findings on the stability of brain AC [2].

Assay of cAMP

Cyclic AMP formed by adenylyl cyclase (AC) activity was assay by the Gilman method [5] modified according to a saturation assay procedure of Brown et al. [9] according to publish specifications [6] but with cyclic [^{3}H]AMP increased to 20,000 c.p.m./assay. The AC membrane preparations from fat cells show an ATPase activity of only 0.1 nmol P_i/h/µg-protein [2], using the assay procedure reported under the reference list [10].

The reaction mixture contained within a final 0.3 ml volume, 50 µl of a standard solution of cyclic adenosine monophosphate (cAMP) or 5µl samples from the reaction assay mixtures or 50 µl from the corresponding zero time blanks, 5 µmol sodium acetate buffer (pH 4.0), 40 µg bovine serum albumin. The assay reactants were added to the reaction mixture: 1.25pmol of cyclic [^{3}H] AMP (NH^{4+} salt containing 8000 cpm). 10 µmol of EDTA pH 4 (which suppressed interference from divalent metals and pH=4.0) and 15 to 30 µg of binding protein from fresh bovine muscle according to the method of Miyamoto et al. (1969) [11]. The amount of binding protein was adjusted to bind about 40% the added cyclic [^{3}H] AMP. The cAMP assay mixture was incubated at 4°C for 1 h, then 3 ml of 20 mM-potassium phosphate buffer (pH=6) were added. The final mixture was filtered through a type HA Millipore filter (0.45 µm pore size) in a 3025 Millipore sampling manifold and vacuum pressure of 20 psi (pounds per square inch), the wells were rinsed with an additional 2ml of the same buffer.

After 5 min the filters were removed, placed in vials with 10ml of Bray's (1960) [12] scintillator fluid and the radioactivity was measured in a Beckman LS-100 scintillation counter. The results (cpm) from zero time blanks and the reaction assay mixtures were corrected for the amount of radioactivity retained by the Millipore filter, in the presence of binding protein. Results expressed as pmol of cAMP formed/h/µg of protein determined by the method of *Lowry* et al. (1931).

Results were calculated by formula, (C- b)/(T- b), C: c.p.m. from the assay of the enzyme activity, T: c.p.m. from total binding (zero-time blanks) and b: c.p.m. in the millipore filters (moisture blanks). The latter values were converted into values of cAMP by using their equivalent intercept within standard curves obtained with a slope of -1. Calibration curves are linear for Log % of C- b/T- b versus log cAMP without interfering ATP or other reagents.

Only values of Log % between 0.10 and 0.85 were used in the calculations because of the increasing error found at the upper and lower limits of the linear plot. Results mean values of triplicates or more: nmol of cAMP formed per hour per milligram of protein. The standard error of the mean (SEM) did not exceed 6% [8]. The cAMP assay was repeated either using samples larger than 50μl of the reaction assay mixtures and their corresponding zero time blanks, or after diluting both to obtain values of Y within indicated limits. The use of zero time blanks to determine individual 100% binding for each of the incubation test mixtures, minimizes the interference by a high concentration of ATP, on the cAMP assay. Also prevents the interference related to a change in the reagent used in the reaction mixtures.

This procedure yields linear standard curves under the same conditions used for zero time blanks (controls).

Purified Rat Liver IRTK Preparation

Soluble insulin receptor was prepared from rat liver membranes as described by Klein et al. [13, 14].

Insulin Receptor Tyrosine Kinase Assay

Partially purified insulin receptor tyrosine kinase (IRTK) was assayed for kinase activity using the peptide phosphorylation procedure of Pike et al. [15].

The preparation containing from 1 to 3 μg of protein according to requirements were incubated for 15min at 23°C in a final volume (25μl) of a reaction mixture containing at final concentrations: 20 mM Hepes (pH 7.5), 1.0 μM insulin, 5% glycerol, 30 mM NaCl, 5 mM $MnCl_2$, ionic Mn^{2+} was maintained constant at 5 mM, 100 μM sodium orthovanadate, and 12 mM p-nitrophenylphosphate. [γ-^{32}P]ATP was added at the final concentrations required and the incubations were continued for 5 min at 23°C. Val^5 angiotensin II was added and the reaction mixtures were incubated for 5 min at

23°C a time at which the rate of Val^5-angiotensin II phosphorylation by IRTK is linear.

Data Analysis

Steady-state velocity equations for a two-substrate sequential mechanism [16], the lines of the double-reciprocal plots were fitted by linear regression. Results expressed as pmol/min/mg.

Chapter 2

The Biological Fundamentals of the Adrenergic System of Brain

Catecholamine Synthesis

Tyrosine by the action of tyrosine hydroxylase forms DOPA (3, 4-dihydroxyphenylalanine). DOPA decarboxylase in the cytoplasm of presynaptic terminals forms dopamine, which is concentrated in vesicles. By the action of dopamine-β-hydroxylase becomes noradrenaline (NA), the final neurotransmitter at the postganglionic neurons of the sympathetic central nerve system. The pathway in the adrenal medulla, by action of PNMT (phenylethanolamine N-methyl transferase), produces adrenaline to be released into blood.

Adrenergic receptors respond to adrenaline, or to NA, or to both. Epinephrine (adrenaline) reacts with both: α- and β-adrenoreceptors, causing vasoconstriction and vasodilation, respectively. Although α receptors are less sensitive to adrenaline, when activated, they override the vasodilation mediated by β-adrenoreceptors because there are more peripheral α1 receptors than β-adrenoreceptors. The result is that high levels of circulating adrenaline cause vasoconstriction. At lower levels of circulating adrenaline, β-adrenoreceptor stimulation dominates, producing vasodilation followed by decrease of peripheral vascular resistance.

It has cardiac ionotropic and chronotropic effects and vasopressor actions on the vasculature. β-adrenergic effects are more pronounced at low doses and

α_1- adrenergic effects at higher doses. It plays a central role in short-term stress reaction, eliciting the fight-or-flight response [17].

The α-adrenergic has greater affinity for NA and the β-adrenergic for adrenaline. There is cross reactivity, and both receptors could respond to high concentrations of both agonists.

NA widens pupils and the air passages of the lungs, contracts and narrows the walls of the peripheral blood vessels by acting on the muscle-located α-receptors [18].

Adrenaline [19] and stress-induced cardiomyopathies [20] redirects blood to essential organs, like heart and brain, but by increasing the heart rate risks increases much the blood pressure and affect the heart rhythm [21].

Central NA pathways are essential in the modulation of mesencephalic dopamine neurons, and also by it mediate behavioral effects induced by methamphetamine.

Neurogenesis

Growth factors, known as neurotrophins, induce transcription process by pluripotent stem cells (iPSCs), involved in the neuronal differentiation processes. The developmental Notch signaling pathway promotes proliferative signaling during neurogenesis. The notch receptor is a protein constituted by a large hetero-oligomer extracellular region. It becomes associated by a Ca^{2+}-dependent coordinative bonding to a single transmembrane-pass, ending into a small intracellular region. The signaling activity is inhibited by Numb to promote neural differentiation [22].

Dysregulation of neurotrophin secretion and function can cause memory impairments and reduced nociception/harmful stimuli (pain). Neuronal differentiation via iPSCs is currently under intense study, as this field may yield therapeutic techniques for spinal cord injuries as well as nervous system diseases.

Wnt signal transduction pathways involve proteins that pass signals from the outside through surface receptors into the cellular inside. Their function has been characterized as the canonical, the noncanonical planer cell polarity and noncanonical Wnt/calcium pathways. These three are activated by the binding Wnt –protein ligand to the *Frizzled* receptor passing signaling to the protein *Dishevelled* inside the cell. The canonical Wnt pathway regulates gene transcription. The planar cell polarity pathway modulates the cytoskeleton, shaping the cell. The noncanonical Wnt/calcium regulates the cell endogenous Ca^{2+} inside the cell. Wnt signaling is characterized by paracrine: nearby cell to cell, or autocrine at the same-cell communications.

The highly conserved by evolution Protein and components of the Wnt are expressed from early development to adult life. These participate in the remodeling of pre- and post-synaptic connectivity, in order to modulate synaptic function. *In vivo* studies indicate a neuroprotective role in Alzheimer's disease.

The deficiency of Wnt proteins had been related to neurological abnormalities and behavioral diseases, by altering the synaptic function at stages where the phenotype has not yet established itself, but cognitive process have started to decline [23].

The process of neurogenesis includes cellular migration to the appropriate brain area, and the growth of dendritic and axonal contacts between neurons.

Inborn neurological disorders result from dysregulation of embryonic neurogenesis, in adulthood stress mechanisms linked to neurogenesis lead to manifestation of psychological dysfunctions.

Presynaptic

Neurosecretory peptides and some neurotransmitters are found in the 100nm large dense core vesicles (DCVs), within the cytoplasm surrounding the small pleiomorphic vesicles within inhibitory synapses. In the latter, the neurotransmitters γ-aminobutyric acid (GABA) or glycine could be found in inhibitory presynaptic boutons. Inhibitory synapses are abundant at proximal dendritic zones and neuronal soma. Inhibitory synapses are located at the axonal hillock, where activation of a small number of inhibitory synapses occurs on the necks of some dendritic spines to regulate axonal firing. In these, can prevent excitatory activation depending on their frequency and the specific circuit involved [24, 25, 26, 27, 28].

80nm small DCVs not only differentiate in size from the larger ones, but also because only 1-10 vesicles occur in the axon boutons of about 20% of mature presynaptic axons. The small DCVs are active zone transporters or local source of new presynaptic active zones [29], involved in rapid synapse formation.

Postsynaptic

Different cell types in the brain could be characterized by differentiable postsynaptic structure. Neighboring dendritic spines on a single hippocampal dendritic segment can vary more than 10-fold in their dimensions. The size of the dendritic spine and its synapse correlates with the number of vesicles in the presynaptic boutons for many brain regions. Neurons that are no longer needed during the developmental process undergo death (neuronal apoptosis), which inhibited by neurotrophins, which activate p75 receptors to express apoptotic genes, also neurotrophins activate the p75 or Trk family of receptor tyrosine kinases activate genes involved in cell growth.

Transversal Media Cut of Brain

Caudal noradrenaline (NA) neurons are separated from the locus coeruleus (LC) complex and located in the A1 and A2 areas [30] of the medullary ventrolateral reticular formation. These neurons restrict target areas and regulate autonomic functions subject to neuroendocrine control [31].

LC is located bilaterally within the lateral pontine central grey, and includes catecholamine neurons close to the brachium conjunctivum and within central tegmental regions of the pontine gray matter [32, 33]. The dorsomedial compartment of LC contains densely packed neurons but these are dispersed within the the ventrolateral compartment [32]. There are 60,000 neurons in the adult human which decrease to 40,000 during ageing [34, 35].

LC neurons diameter is 15-25 μm for the small spindly tufts of dendrites found within the ventral LC complex [32]. The medium sizes 35-45 μm of the multipolar cells are generally round, with only a few dendrites located in the rostral part of the nucleus, which often extend into surrounding.

The brain stem has ascending nerves that control sensory functions, while the descending ones, are the motor tracts. Neurons in the spinal dorsal horn (brain stem) process sensory relay (information), receives monosynaptic input from mechanoreceptors (A-fibers) and polysynaptic input from nociceptors (C-fibers). Ca^{2+} increases GLU and substance-P for presynaptic release postsynaptically Glu-receptors α-amino-3-hydroxy-5-methylisoxazole-4-propionate (AMPA) induces a fast transient the depolarization of the membrane preventing the Mg^{2+}, blocking of the *N*-methyl-D-aspartate ((−)NMDA)-type receptors.

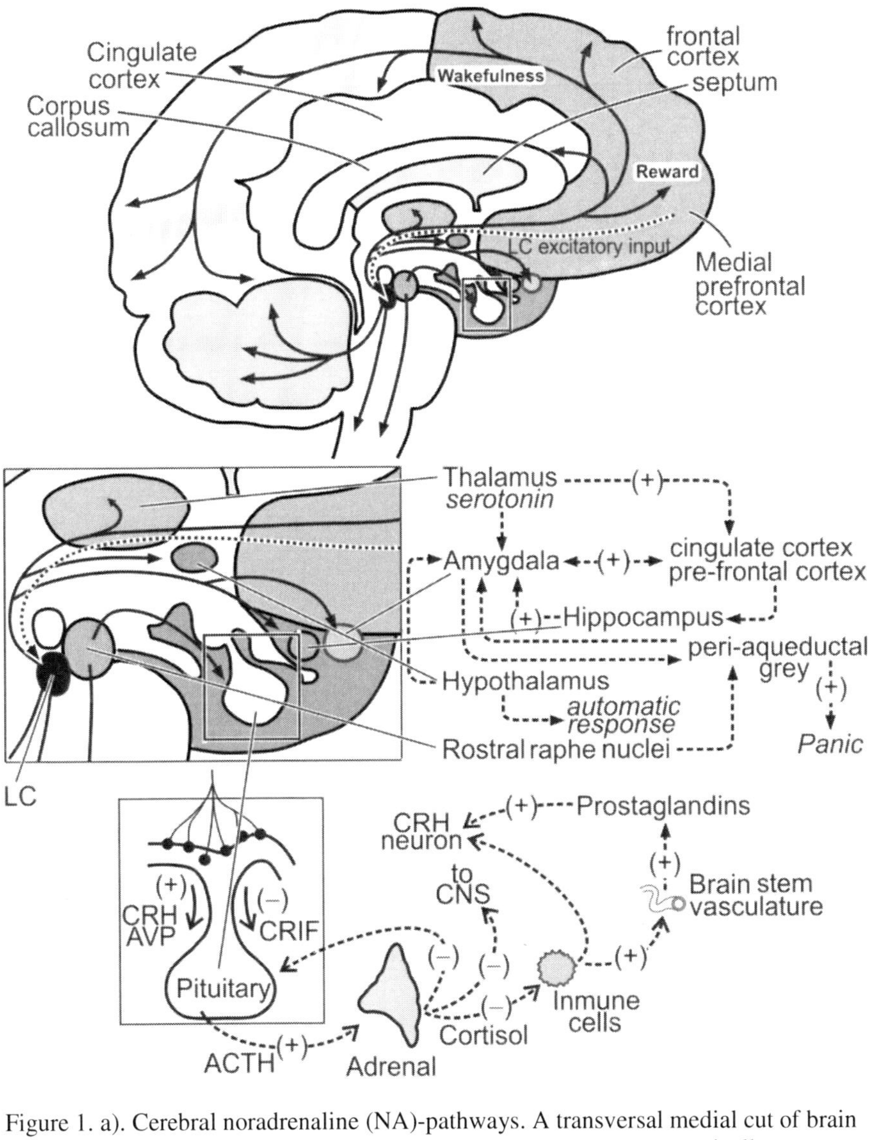

Figure 1. a). Cerebral noradrenaline (NA)-pathways. A transversal medial cut of brain illustrates the locus coeruleus NA-releasing efferent neurons, presynaptically activating postsynaptically located adrenergic receptors. NA binding activates various subtypes of adrenergic receptors for a differential role of signaling pathways, which functionally operate by inducing synthesis of second messengers. The effect of Mg^{2+} is to up-regulate adenylyl cyclase (AC) and cyclic adenosine monophosphate (cAMP) formation. Subtype $\alpha 2$ adrenergic receptors ($\alpha 2$-AR) coupled to Gi/Go inhibit presynaptically by a decrease in the Ca^{2+}-dependent release of NA containing vesicles, down-regulating AC and inhibiting Camp formation.

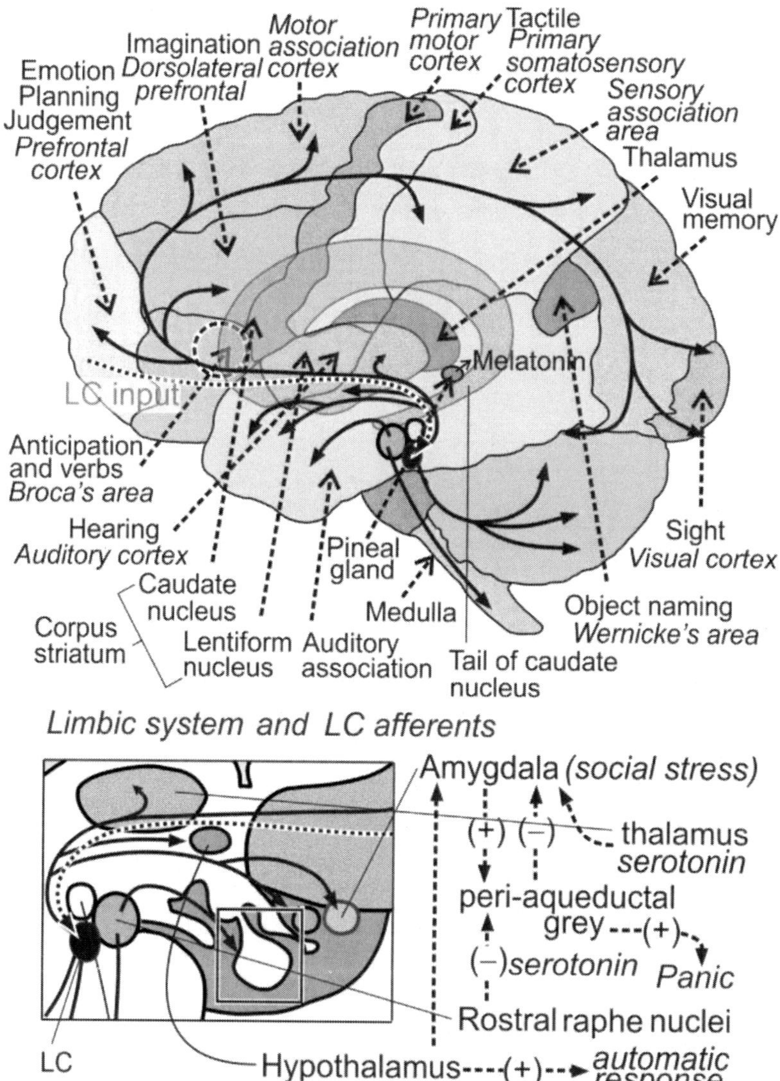

Figure 1. b). Regions of the human brain. Emotion originated in the limbic system including amygdala for behavior frame of mind, memory, pleasure and addiction. Broca's area: language production anticipation and verbs. Wernicke's area: understanding and compression object naming. Orbital frontal area functions for rezoning guided by emotion, which involves decisions taking structures on planning and attention. Sound perception is a reward area link to release of dopamine, which is also secreted with food and sex. The hippocampus (two hippocampi) consolidates information from short-term memory to long-term memory and spatial navigation.

The left amygdala has been linked to social anxiety, obsessive and compulsive disorders, and post-traumatic stress, as well as more broadly to separation and general anxiety. For motivation specifically, it matters which pathway takes dopamine. The mesolimbic pathway, which originates in the middle of the brain and branches to various places like the cerebral cortex, is the most important reward pathway in the brain. Visual association area for discerning shapes and location involves 30 areas different mobilizing up to 25% of brain activity.

Other brain areas release the neurotransmitters: γ-aminobutyric acid (GABA), glutamate (Glu), hypocretin, orexin and corticotropin factor (CRF).

The pH-sensitive voltage-gated Ca^{2+}-channels (VGCC) are inhibited in the presynaptic membrane of retinal ribbon synapses [46] [47] [48].

Instead of lymph, the brain is bathed in cerebrospinal fluid (CSF).

The α_2 receptor mediates synaptic transmission in pre- and postsynaptic nerve terminals, and are involved in decrease release of acetylcholine, norepinephrine and inhibit the brain NA system. The receptors inhibit [36] lipolysis in adipose tissue, and the insulin release in pancreas. The α_2 receptors are involve in induction of glucagon release from pancreas.

Chromaffin cells are postganglionic neurons that directly secrete epinephrine into the blood stream as part of the sympathoadrenal system. The postganglionic neurons of the sympathetic nervous system secrete norepinephrine onto target tissues.

LC-Noradrenaline-Dependent Activation of Emotional Responsiveness

Locus coeruleus (LC) is located in the posterior area of the rostral pons, in the lateral floor of the fourth ventricle. Inside the neurons, granules of neuromelanin formed by polymerization of noradrenaline (NA) confer a blue color.

The axons of LC neurons reach cerebellum, brain stem and spinal cord, hypothalamus, thalamic relay nuclei, hippocampus and amygdala, basal telencephalon, tectum and cortex (figure 1.a). Ca^{2+} triggers that NA vesicles presynaptically release the neurotransmitter which has an excitatory effect on most of the brain, priming the brain's neurons to be activated by stimuli to mediate arousal.

It is involved in several neural pathways responsible for somatic and emotional processes, mood alterations, such as pain and depression, and the physiology and pathology modulation of sleep-waking cycles, movement disorders convulsive seizures. The alterations of LC activity result in complex patterns of neuronal function. *Via* it's the regulation of arousal and autonomic activity [37].

LC is a major wakefulness-promoting nucleus, resulting from dense excitatory projections to the majority of the cerebral cortex. Activation of the LC, thus results in the enhancement of alertness through the innervation of cholinergic neurons of the basal forebrain, cortically-projecting neurons of the thalamus. This effect extends to serotoninergic neurons of the dorsal raphe and cholinergic neurons of the pedunculopontine and laterodorsal tegmental nucleus, and substantial inhibitory projections to sleep-promoting GABAergic neurons, which produce γ-aminobutyric acid (GABA) of the basal forebrain and ventrolateral preoptic area.

The importance of the LC in controlling autonomic function results from both direct projections to the spinal cord and projections to autonomic nuclei. This includes the dorsal motor nucleus of the vagus, the nucleus ambiguus, the rostral ventrolateral medulla, the Edinger-Westphal nucleus, the caudal raphe, the salivatory nuclei, the paraventricular nucleus, and the amygdala. LC activation produces an increase in sympathetic activity and a decrease in parasympathetic activity *via* these projections.

LC is the major noradrenergic nucleus of the brain, giving rise to fibers innervating extensive areas throughout its neuroaxis.

LC in adults contains 2.5 to 5×10^4 neurons that range in size between 30,000 and 60,000 μm^3 [38]. The figure 1.a) and 1.b) illustrate that their long axons branch to innervate the brain stem, spinal cord, cerebellum, hypothalamus, thalamic relay nuclei, neocortex and amygdala. Specifically the neuroanatomy of memory involves the hippocampus, the striatum, or the mammillary bodies.

LC-axons have about 10^3 synaptic connections per neuron and through a Ca^{2+}-dependent mechanism can release NA at all their termini. Emotional charged events have been related to the activation of LC-release of noradrenalin [39]. Thus, allows inferring that an LC-modulated perception and limited in capacity and duration, fits into short-term defined parameters of memory.

The activity of LC neurons has been extensively investigated in a variety of behavioral states. In fact, this NA-containing nucleus modulates many physiological and pathological conditions. This includes the sleep-waking cycle, movement disorders, mood alterations, convulsive seizures, and the effects of drugs such as psychostimulants and opioids.

Central NA-pathways modulate the behavioral and neurotoxic effects produced by the psychostimulant methamphetamine. It also integrates the modulation of the activity of mesencephalic dopamine (DA) neurons. NA mediates some behavioral effects induced by methamphetamine. NA modulation of DA release is pivotal for methamphetamine-induced behavioral states and neurotoxicity. These interactions are discussed elsewhere on the basis of the state of the art of the functional neuroanatomy of central NA- and DA systems. Other authors emphasise the role of brain sites, possessing a remarkable overlapping of both neurotransmitters.

LC-neurons are in a much smaller number than the 10^{11} glutaminergic neurons. Neuronal engineering may require inhibition of the glutaminergic neurons to restrict swelling. Mg^{2+} stimulates noradrenergic response at the time that the Mg^{2+}-dependent vasodilation of the arterioles, allows to increase by 10-fold the influx volume of blood. Swelling could be restricted to rather small areas thanks to small relative area participation of LC neuronal axons. During sleep the brain shrinks [40], which allow to postulate a decreased in the stimulated AC response.

The ionotropic N-methyl-D-aspartate (NMDA) type receptors of the glutaminergic neurons could be characterized by Mg^{2+} blocking the capability of glutamate (Glu) to activate their uptake of Ca^{2+}.

Glu-receptors α-amino-3-hydroxy-5-methylisoxazole-4-propionate (AMPA) induces a fast transient depolarization of the membrane, preventing the Mg^{2+} blocking of the NMDA receptors.

The overall Mg^{2+} vs Ca^{2+} effect is to separately activate opposite neuronal responses between LC neurons and glutaminergic ones, facilitating synchronization and alternative dominance. The latter, could became part of a diurnal vs sleep repackage of information.

This in theory may isolate the LC-neurons performance from a *cacophonic* interference by the more numerous glutaminergic ones.

Synchrony between Mg^{2+} vs Ca^{2+} alternative stimulation could restrict the Mg^{2+} promoted-swelling of the arterioles to only those in contact with the adrenergic synapses. This effect by an arteriole up-stimulation of the NA-responsiveness of the postsynaptic located adenylyl cyclase (AC) could fine-tune the stimulus with a target neuronal circuit.

Emotional-charged perception could be compatible with an increment of the simultaneous connectivity between a relatively small numbers of neurons.

In the mammalian brain LC is one of the nuclei with non-deterministic properties could be involved in estimation of environmental changes. The brain could rely in recent fight-or-flight response outcomes to find survival strategies, and therefore evolution could acquire the tendency to develop controls of neuronal plasticity.

LC activation measured by means of pupil diameter connect anterior cingulate cortex (ACC) and LC response appear to be connected for environmental volatility. Estimation of learning rates could be model as a computational tool [41].

Figure 1.a) and 1.b) illustrate that the axons distribution of LC-neurons could allow a near simultaneous convergence into LC-*episodic* memory. Hence, assigning emotional overtones of anger or fear during perception of danger from the surrounding events conformed by diverse-sensory inputs.

The presynaptic Ca^{2+}-dependent release of NA activating the postsynaptic located NA-receptors of AC could mediate integration of the stimulation of specific sensory neurons into specific neuronal circuits.

The LC axonal release of NA vesicles shows an up-feedback inhibited by the presence of α_2-adrenergic receptors within this neuron itself. At the presynaptic junction NA is reuptaked through a co-transported process coupled to Na/Cl. Cocaine blocks the reuptake of catecholamines.

NA inhibited the dopamine uptake and stimulatory action over glucose-induced insulin secretion. Effect reversed by the dopamine D2 receptor to the addition to the antagonists like the antipsychotic butaclamol [42, 43].

Overstimulation of a specific area could produce at the respective synaptic clefts a large discharge of NA, which if not promptly reuptaked, results on an acute panic attack. The process is inhibited by the neurotoxins responsible for tetanus and botulism [44].

NA may affect the heart rhythm, under non-emergency situations: linezolid, beta-blockers such as propranolol, monoamine oxidase inhibitors (MAOIs) are antidepressants such as phenelzine, rasagiline or selegiline for Parkinson's disease, general anaesthetics, and tricyclic antidepressants such as amitriptyline.

NA can reduce the blood supply to the placenta, but it is poorly absorbed from the gut and therefore do not pass into breast milk.

The NA-mediated vasoconstriction optimizes spatial and temporal focusing of the hyperemic response optimizes neurovascular coupling with a 6-fold decrease in the disparity between blood volume and oxygen demand. This enhances synchronization of the redistribution of blood delivery to subsequently active regions.

Brain Regions Afferent Signaling to Postsynaptic LC-Neurons

The norepinephrine from the locus coeruleus (LC) has an excitatory effect on most of the brain, mediating arousal and priming the brain's neurons to be activated by stimuli.

LC axons innervate the spinal cord, the brain stem, hypothalamus, the thalamic relay nuclei, the amygdala, cerebellum the basal telencephalon, and the cortex. The LC receives afferents from the hypothalamus, which is stimulated by glucose level to regulate metabolic homeostasis within brain regions.

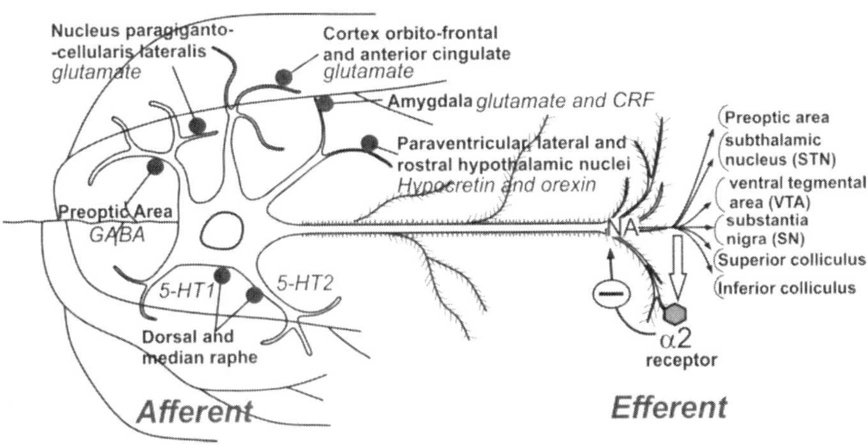

Figure 2. LC-noradrenaline (NA) neurons receive inputs from other brain regions. α2-adrenergic, opioid and 5-HT1A receptors, dorsal raphe nucleus (DRN) serotonergic neurons release at the synaptic cleft serotonin (5-HT1) and its metabolite: 5-hydroxyindoleacetic acid (5-H1AA) [45]. NA-release is control by a negative feedback on the α2-receptor at its axonal terminus. GABA (γ-aminobutyric acid) receptors are pH-sensitive. The cingulate gyrus and the amygdala allow stressors to trigger noradrenergic responses and emotional pain.

The cerebellum and afferents from the raphe nuclei also project to the LC, in particular the nucleus raphes pontis and nucleus raphes dorsalis. Also other regions include medial prefrontal cortex, whose connection is constant, excitatory, and increases in strength with raised activity levels in the subject, nucleus paragigantocellularis, which integrates autonomic and environmental stimuli, nucleus prepositus hypoglossi, which is involved in gaze, Lateral hypothalamus releases orexins, excitatory of the LC neurons.

The endorphins could be displaced by morphine from the opioid receptor is antagonist naloxone, could not reverse ligand-inhibition, as it occurs by subsequent administration of the α_2-adrenoceptor antagonist, idazoxan. Venlafaxine an antidepressant with analgesic properties produces inhibition of LC-neurons.

Other brain areas release the neurotransmitters: γ-aminobutyric acid (GABA), glutamate (Glu), hypocretin, orexin and corticotropin factor (CRF).

The pH-sensitive voltage-gated Ca^{2+}-channels (VGCC) are inhibited in the presynaptic membrane of retinal ribbon synapses [46, 47, 48].

Instead of lymph, the brain is bathed in cerebrospinal fluid (CSF).

Serotonin

Serotonin (5-hydroxytryptamine (5-HT)), biochemically derived from tryptophan, 90% of total serotonin is primarily located enterochromaffin cells in the gastrointestinal tract (GI tract). It is metabolized mainly to 5-HIAA, chiefly by the liver. Metabolism involves first oxidation by monoamine oxidase to the corresponding aldehyde. This is followed by oxidation by aldehyde dehydrogenase to 5-HIAA, the indole acetic acid derivative.

5-HT regulates intestinal movements [49], platelets and the central nervous system (CNS) of animals, including humans. The other 10% is synthesized in serotonergic neurons of the CNS.

Serotonin also has some cognitive functions, including memory and learning. Modulation of serotonin at synapses is thought to be a major action of several classes of pharmacological antidepressants. It is popularly thought to be a contributor to feelings of well-being and happiness [50]. The overall effect will promote an individual feeling of emotional betterment, such that in some cases could lead to eating disorders like binging on sweets, starches or carbohydrates in order to reach a feeling of well-being. Stimulation of certain brain areas result in increased affluence of blood and neuronal uptake of plasma/glucose could be regarded as a reward signaling, which would increase the activity of the favored neurons.

Low serotonin levels could contribute to a person's sense of depression, low 5-HT has been found in suicidal and in particularly aggressive persons.

The Amygdala

The amygdala sends impulses to the locus coeruleus (LC) hypothalamus, the dorsomedial thalamus, the facial nerve, the ventral tegmental area and the laterodorsal tegmental nucleus the thalamic reticular nucleus, the nuclei of the trigeminal nerve, and the medial nucleus involved in smell and pheromone recognition with input from the olfactory bulb and olfactory cortex [51, 52, 53]. The amygdala is associated with the feeling of fear removal disinhibit feelings of lust and curiosity.

The amygdala itself is a highly complex collection of nuclei, so it could conceivably support different emotions in different brain areas to manifest anger and fear.

Adult and adolescent bipolar patients characterize for amygdala and somewhat smaller hippocampal smaller volumes [54]. Monkey mothers who had amygdala damage showed a reduction in maternal behaviors.

Neuroimaging technology shows the amygdala has a substantial role in mental states and relates to many psychological disorders, children with anxiety disorders tend to have a smaller left amygdala. The left amygdala has been linked to social anxiety, obsessive and compulsive disorders, and post-traumatic stress, as well as more broadly to separation and general anxiety. Borderline personality disorder showed significantly greater left amygdala activity. Individuals with psychopathy show reduced autonomic responses, relative to comparison individuals to instructed fear cues [55]. Hyperactivity the amygdala was shown when patients perceive themselves in frightening situations.

Social phobia showed a correlation with increased response in the amygdala. Depressed patients showed exaggerated left amygdala activity when interpreting emotions like fearful faces, the effect disappear under administration of antidepressant [56], but bipolar patients reacted differently.

Cerebral Cortex

Processing of visual information, emotional experience in humans may not be entirely sub-cerebral as reflected by brain right/left dichotomy and the emotional effects of prefrontal lobotomy. The cerebral cortex side is capable of semantic memory, the ability to learn facts and relationships mediated by changes in the connections between cells, related by their conducting of specific types of information [57]. The evidence indicates multiple centers of feeling in the brain, associated with will, self-image, conscience, etc.

Anterior Temporal Lobe

Produced noticeable changes, including overreaction to all objects, hypoemotionality, loss of fear, hypersexuality, and hyperorality, in which inappropriate objects are placed in the mouth.

Nucleus accumbens and the septal nuclei concentrate pure pleasure. Globus pallidus produces an experience of joy Cingulate cortex interests qualify as feelings, Orbitofrontal cortex Guilt, anxiety and paranoia.

Hypothalamus

A daily photic signaling to the hypothalamic suprachiasmatic nucleus (SCN) became the central oscillator for the appropriate phasing of the various biological rhythms.

This brain region has become a center for the filings of hunger and thirst.

The ventromedial hypothalamus (VMH) has been shown to respond to catecholamine released into the glucose-sensing region, through a counter-regulatory response to hypoglycemia by stimulation of the β_2-adrenergic-receptor (B2AR) [58].

Fibroblast growth factor 21 (FGF 21), a hepatokine, increases systemic glucocorticoid levels and alters circadian behavior through expression of the β-Klotho gene in the suprachiasmatic nucleus of the hypothalamus and the dorsal vagal complex of the hind brain [59].

Arginine Vasopressin

Arginine vasopressin or antidiuretic hormone (ADH), it is derived from a preprohormone precursor, synthesized in the hypothalamus and stored in vesicles at the posterior pituitary. It has a very short half-life between 16-24 minutes [60]. ADH is a neurohypophysial peptide hormone that increases water permeability of the kidney's collecting duct and distal convoluted tubule by inducing translocation of aquaporin-CD water channels, increase water reabsorption in the collecting ducts of the kidney nephron [61, 62]. Promote homeostasis by regulation of glucose and salts. It plays an important role in social behavior, sexual motivation and pair bonding, and maternal responses to stress.

Oxytocin

Oxytocin produced by the hypothalamus, stored and secreted by the posterior pituitary gland acts as a neurohypophysial/neuromodulator of intimacy, sex and reproduction. It is released in large amounts after distension of the cervix and uterus during labor, facilitating birth, maternal bonding, and, after stimulation of the nipples, lactation [63].

In behaviors acts as *bonding-hormone* includes maternal, orgasm, social recognition, pair bonding, anxiety, and [64] ethnocentric behavior, by incorporating empathy and trust of in-groups debilitating suspicion and rejection of outsiders [65].

A defective oxytocin receptor gene has been associated with maladaptive social traits such as aggressive behavior [66].

Hippocampus

Hippocampus contains two main interlocking parts: Ammon's horn and the dentate gyrus. In Alzheimer's disease, the hippocampus is one of the first regions of the brain to suffer damage.

Hippocampus is believed to be involved in spatial learning and declarative learning. A very short attention span could not adequately sustain survival and therefore evolution lead to a hippocampus capable to process learning into a pattern for the consolidation of new information even if cannot stored itself.

Pharmacological manipulations in hippocampal slices indicate that glutamate is primarily released through opening of the anion channels of astrocytes.

Since the channel opening could last from seconds to minutes are not required for quick responses.

The actions on the cell can be more widespread than the more localized ionotropic effects.

The neurotransmitter has the potential to activate the metabotropic glutamate receptors (mGluRs) interacting with G proteins. The transduction mechanisms do not only result in ion channels opening but also involve second messengers.

Activation of the glutaminergic receptors can interlink Ca^{2+} channels of the glutaminergic neurons and astrocytes to orchestrate the Ca^{2+} activation of Na^{+}/K^{+}-ATPase to allow the uptake of extracellular K^{+}. The effect contributes to a nerve impulse better performance on the signal-to-noise ratio.

This finding appears to relate the brain sleeping state, with adrenergic receptor inhibition by Ca^{2+}.

The hippocampus is also important into long-term explicit memory, receives input from secondary and tertiary sensory areas, process information send by the cortex to different regions in the brain. Damage causes problems with memory storage [67].

The mammalian hippocampal formation provides neuronal representations of environmental location [68]. The majority of cells in medial entorhinal cortex and parasubiculum show spatially periodic firing patterns. Grid cells exhibit hexagonal symmetry and form an important subset of this more general class [69].

Facilitation of spontaneous sharp-wave ripple complexes (SPW-Rs) in hippocampal slices by increasing gap-junction or γ-aminobutyric acid-A ($GABA_{(A)}$)-mediated axon depolarization was shown to reduce of synaptic strength, and electrical stimulation evoked in axons, long-lasting synaptic depression. This synaptic depression requires L-type Ca^{2+} channel activation. Synaptic stimulation delivered antidromic SVT (conduction down the accessory pathway and back up through normal pathway). This form of firing was shown unable to induce potentiation to a long-lasting synaptic strength and was suggested that the effect could contribute to memory consolidation during resting by conferring specificity to subsequent synaptic input [70].

Dreaming allows electrical impulses to be propagated like waves, which induce in the hippocampus cells to activate the dromic direction which is contrary to the habitual antidromic. Thus, sends electrical signaling from their axons to the neuron's own dendrites. Is has been assumed that the effect could produce space in were to store the new memories for the awakening person. This methodology developed has been shown to improve Parkinson symptomatology.

Inhibitory synapses can be interspersed among excitatory synapses, about 1 inhibitory per 10-20 excitatory synapses, along a dendrite.

The *striatum* and the hippocampus have been shown by functional magnetic resonance imaging (FMRI) to interact cooperatively to support episodic memory formation [71]. New long-lasting details of specific events could be lost to people that show amnesia, a dysfunction that can be related to hippocampus damage [72].

Small dense core vesicles (DCVs) are a local source of new presynaptic active zones [29, 73], during rapid synapse formation in the mature neurons supporting a role of a local source of new presynaptic active-zones [74].

96% of the vesicle-containing boutons within the mature neurons interact with one or more postsynaptic partners. Single-synapses are about 75%, 21% multisynaptic and only 4% are non-synaptic boutons [75, 76]. The number of multisynaptic boutons increases in rapid synaptogenesis [77]. The multisynaptic dendritic spines receiving the input of more than one several presynaptic axon occur rather frecuentrly. The non-synaptic boutons may be a source of presynaptic one for rapid synaptogenesis.

Neighboring dendritic spines on a single hippocampal dendritic segment can vary more than 10-fold in their dimensions. Large highly branched dendritic spines are the postsynaptic partners of mossy fibers in this area.

Amnesia can result from extensive damage to regions of the medial temporal lobe: hippocampus, dentate gyrus, subiculum, amygdala, and the parahippocampal, entorhinal, and perirhinal cortices. Regions of the midline diencephalic region: dorsomedial nucleus of the thalamus and the mammillary bodies of the hypothalamus [78, 79].

There are many sorts of amnesia, and by studying their different forms, it has become possible to observe apparent defects in individual sub-systems of the brain's memory systems, and thus hypothesize their function in the normally working brain. Other neurological disorders such as Alzheimer's disease and Parkinson's disease [80] can also affect memory and cognition. Hyperthymesia, or hyperthymesic syndrome, is a disorder that affects an individual's autobiographical memory, essentially meaning that they cannot forget small details that otherwise would not be stored [81]. Korsakoff's syndrome, also known as Korsakoff's psychosis, amnesic-confabulatory syndrome, is an organic brain disease that adversely affects memory.

Cerebellar Cortex

Parallel fiber axons synapse with dendritic spines of Purkinje cell spiny branchlets, climbing fiber axons that synapse along the proximal dendrite of these cells, have larger and more irregularly shaped boutons than the parallel axons.

The dendritic spines distributed along the same presynaptic axon vary greatly in their dimensions also the ones along the spiny branchlets of cerebellar Purkinje cells vary in their size. These cerebellar spines can be branched but the head of each branch on the spine is club-shaped.

Memory Modulation

Encoding or registration: receiving, processing and combining of recently obtained information.

Storage: recording in permanent form encoded information.

Retrieval: calling back the stored information through cue processes.

Forgetting: involves seven types, characterized as prescriptive forgetting, formation of new identity, structural amnesia, annulment, planned obsolescence, repressive erasure and humiliated silence. Forgetting involves activation of the prefrontal hippocampus in a way similar to movement inhibition. If an event could not be retrieved could mean imperfect encoding of information. Reviewing information can activate a retrieval which seems to slow the rate of forgetting. Older adults that had less episodes of forgetting events are more active than those who were less active.

It could be expected that a large number of the pathways associated with neuronal activity, could be integrated into the complex process of memory formation and consolidation, by an *in common* coupling by the interdependence of signal and ATP-energy transduction mechanisms [1, 82, 83, 84].

Up-regulation of adenylyl cyclase (AC) by ionic equilibrium in dependence of changes in ATP concentrations, leaded to investigate possible metabolic-ionic feedbacks related to the energy requirement of the noradrenaline (NA)-AC system. Stimulated neurons require O_2 to maintain the supply of ATP for the Na^+-pump. O_2 increases turnover of both electron transport chain and Krebs cycle, which occurs through the increase in dicarboxylic and tricarboxylic acids concentrations in mitochondria. Effect which increases Ca^{2+} retention in mitochondria and up-regulate the Na^+/K^+-ATPase-AC system. The relationship between the rate of aerobic glycolysis and neuronal work was therefore examined, in the context of the signal transduction of tissue networking. The latter, allows a large increment in the rate of glucose and O_2 uptake by stimulated brain areas, even when glucose levels in blood are homeostatically sustained.

Excessive or prolonged with cortisol stress appears to hurt memory storage. Direct injection of cortisol or epinephrine at the level of the amygdala helps the storage of recent experiences.

NA is reuptaked at the presynaptic junction through a co-transported process coupled to Na^+/Cl^-. Inside the neuron is destroyed in the mitochondria by the action of monoamineoxidase (MAO), and inside the cytoplasm by catecholamine-O-methyltransferase (COMT). The resulting metadrenaline and normetadrenaline are subject to the consecutive action of monoamine oxidase and COMT to form HMMA (4-hydroxy-3-methoxymandelic acid). Cocaine blocks the reuptake of catecholamines. Axons usually release at all their termini the same neurotransmitter.

NA activates AC in membrane preparations obtained from rat's cerebral cortex, corpus striatum and hypothalamus [2], activating a second messenger system. Over-exposure by pre-incubation of AC with NA results in that the afterward assay AC activity shows that prolonging exposure to the neurotransmitter inactivate the enzyme

It could be inferred that NA-dependent psychosomatic damage, lead in the course of an evolutionary selection process to a costly protection mechanism. This one turns-off the enzyme rather than allowing a too perdurable permanence of the organism into the fight-or-flight response.

Synaptic remodeling involves physical changes in the neuronal synapse required to recycle receptors and to create new synapses generating pathways, which reinforce the strength of connectivity between neurons to reinforce recall of events to become part of long-term memory (LTM).

The postsynaptic density of receptors within the neuronal membrane represents sites specialized in learning and memory.

The brain recalls immediate events to capture information of "what", "when" and "where" by a short-term memory supported by transient patterns of locus coeruleus (LC) neurons, communicating regions of the dorsolateral prefrontal cortex and the parietal lobe.

Short-term memory appears to rely mostly on acoustic components of a code, but to a lesser extend could use encoding of written text through a visual code.

Short-term memory was adapted by evolution for an *emotional* integrated-recording of multiple perceptions of events, denominated *episodic*, in which vision-auditory-olfactory-touch-gustatory are involved in the nurturing of LTM.

Storage in short-term memory has a limited capacity and duration. LTM is semantically encoded and can store a much greater quantity of information without time limitation. This one is maintained by more stable and permanent changes in neural connectivity, almost widely spread over the brain.

Sleep between training and test improve the consolidation of information, neuroimaging studies discovered activation patterns in the sleeping brain that include rehearsals, which mirror the tasks done over the previous day.

LTM is required for later recall. Memory consolidation requires synaptic plasticity characterized by gene expression and physical changes for construction of new synapses [85, 86].

Immediate-early response genes (IEGs): ARC coding Arc (activity-regulated cytoskeleton-associated protein), CEBPB, PCDH8, CEBPD, CREB1, CREM, EGR1, EGR2, EGR3, EGR4, BDNF, FOS, HOMER1, JUNB, KLF10, MMP9, NFKB1, NFKBIB, NGF, NPTX2, NR4A1, JUN, NTF3, PIM1, PLAT (tPA), RELA, RGS2, RHEB, SRF, TNF, are activated after neuronal events to alter expression. IEGs mediate long-term potentiation (LTP). LTP genes: ADCY1, ADCY8, BDNF, CAMK2A, CAMK2G, CDH2 (N-cadherin), CNR1, GABRA5, GNAI1, GRIA1, GRIA2, GRIN1, GRIN2A, GRIN2B, GRIN2C, GRIN2D, MAPK1, MMP9 (Gelatinase B), NTF4, NTRK2, PLCG1, PPP1CA, PPP1CC, PPP3CA, PRKCA, PRKCG, RAB3A, YWHAQ, processes that enhance synaptic connections participate in learning and memory consolidation.

An opposite response occurs through the expression of long-term depression (LTD) genes: GNAI1, GRIA1, GRIA2, GRIA3, GRIA4, GRIP1, GRM1, GRM2, IGF1, MAPK1, NOS1, NGFR, PPP2CA, PPP3CA, PRKCA, PICK1, PLAT (tPA), PPP1CA, PPP1CC, PPP1R14A (CPI-17), PRKG1, inhibit synaptic connections.

Robotized transcranial magnetic stimulation (TMS) which in response to electric current induced conscious neurons to fire allowed perception of light flashes (phosphenes) established internet-mediated through a brain-computer interfaces (BCI) between a sender's brain to individual receivers, which have the sensation of decoding the 1 and 0 sequence as the transmission of the word *hello* [87].

Unconscious cognition processes learning, perception, memory, thought, and language without awareness.

According to Freud the Consciousness is a center for perception, which directs to an unconscious level, past thoughts and memories.

The actual level of involvement of an unconscious brain during a cognitive process could be subject to differential opinion, but it is undeniable its role to influence thoughts and feelings.

There have also been several experiments, which point to the unconscious mind could be better at decision making than the conscious mind.

Freud assumes that the Unconscious' influence on thoughts can be accessed by meditation, by random association, dream analysis and paying attention to Freudian slips.

Carl Jung's categorized the unconscious into a personal and a collective unconscious. The latter, could be passed-on from one generation into another and shared among all the individuals of a species.

Lacan's linguistic unconscious was not developed as a reference to self, but to a dynamic structure, which could be altered by trauma or brain injury, and explain personality syndromes.

The Role of Amygdala in Emotional Learning and Memory

The hippocampus is a fundamental structure for recognition of face and facts but the amygdala adds the emotional perception. This implicates the brain as functioning between two differentiable registration-event systems. One is for the ordinary and the other for the emotional-charged events. The latter, triggers the LC-network to release NA, which presynaptically activates postsynaptically located receptors, because this are present in the hypothalamus activates the HTPA axis and the adrenals release adrenaline.

The amygdala has been associated with emotional events participating in the formation of memory, adequating the fight-or-flight response to form associations between stimuli and the aversive events. These form a library of unconscious responses in the amygdala, from which fear conditions the sensory stimuli to reach the basolateral complexes and lateral nuclei.

A sustained enhancement between affected neurons [88] endures and could damage the amygdala. Thus, interferes with the long-term potentiation connected to emotional learning [89], the acquisition and expression of Pavlovian fear conditioning. Different nuclei within the amygdala have different functions in appetitive conditioning [90].

The genesis of many fear responses like freezing or fight-or-flight response involves the central nuclei and links to autonomic nervous system into neuroendocrine, changes in blood pressure and heart rate/tachycardia.

In a functional magnetic resonance imaging (FMRI) study, more intensive insula activity was found in expert meditators than in novices. Increased activity in the amygdala following compassion-oriented meditation may contribute to social connectedness [91].

Emotional arousal on the amygdala neurons shows various types of oscillation like theta activity.

Synchronized neuronal events by increasing interactions between neocortical storage sites and temporal lobe structures involved in declarative memory [92] could promote the memory retention requiring synaptic plasticity.

Rorschach/blot 03 by its random shape elicits a larger number of responses by an artistic population and the test had been linked to large-sized amygdala [93].

Information derived from learning event does not consolidate into memory instantaneously. The amygdala is not itself a long-term memory and learning can occur without it, but it does regulate memory consolidation in other brain regions [94].

The amygdala especially the basolateral nuclei are involved in emotional arousal influencing the strength or subsequent persistence for an event to became consolidated into memory. Similar effect is obtained by administration of stress hormone.

Classical conditioning tasks resulting in inhibitory avoidance show that drugs, which activates the amygdala, result in a better memorization of the training task and viceversa, drugs that inhibit this region result in impaired memory.

Emotionally-arousing information correlates with increase in the amygdala activity at the encoding and retention times.

The Role of Glucose as a Reward Mechanism for the Consolidation of Cognitive Circuits

Hemoglobin histidine R-groups were shown to conform a dynamic binding site for Mg^{2+} or Zn^{2+}, [95] other authors applied the same characterization under the denomination of Zn^{2+}-fingers. The participant histidines provide turnover to the chelating activity of the metal binding site, by increasing affinity of Hb for the mutual inclusion at the allosteric separated sites for Mg^{2+}/Zn^{2+} and for O_2, couple to an increase of 60 water molecules in the hydration shell of oxyHb. The protein by allowing Cα rotation to the amphoteric R-histidines releases the chelate-metal from the His groups, which turn to participative in the conformation of a separated site. The latter by binding of 2,3-DPG decreases affinity for O_2 and Mg^{2+} leading to their release at target tissue which for brain correspond to synaptic clefts.

These processes interact through the reordering of H-bonds for simultaneous energy- and signal-transduction [96] functionality of electrogenic membranes.

Several research groups using positron emission tomography detected the topographical description dynamics by the response to 2-deoxy-2-(18F)fluoro-D-glucose acting for deep brain stimuli [97]. Concomitantly, the metabolic activation pattern found that glucose uptake and its metabolism precedes for same time the initiation of O_2 consumption.

The adrenal gland releases adrenaline into blood which induces metabolic shifts, increasing the production of glucose a metabolic reward for neuronal action. Hence, whereas the blood-brain barrier prevents that the adrenaline secreted in blood could reach the brain itself, glucose does. Its uptake conditions neurons to crave for glucose, an excitatory molecule, and therefore a reward to memory, which acts to trigger a rapid response of a targeted emotional circuit.

This indicates that in brain the formation of the glucose transporter (GLUT3)-glucose-complex and turnover has a signaling role. Progress of the reaction may involve that binding of glucose, lead to the release of water molecules from the hydration shell of GLUT3 to the outside of the membrane. This adds to the tendency of the complex to transverse the hydrophobic environment in the membrane, releasing glucose into the inside. Un-complexes GLUT3 could acquire the tendency to move seeking the exterior hydric environment to reuptake its hydration shell.

Examination of evolution paths, suggests links between reward mechanisms and cognitive development. Mg^{2+}-dependent responsiveness to noradrenaline of adenylyl cyclase (AC) increases cyclic adenosine monophosphate (cAMP) production. The latter, has been linked to long-term emotional cognition and learning. The release of glucose into the synaptic cleft could nurture a tropism mechanism for neuronal dendrites, seeking an enhanced reward by participating into newly develop cognitive circuits. A

GLUT3-glucose complex could displace across neuronal membranes involves vibrational patterns, which may code-decode for a reward of sensory neurons evoking the neuronal impulses, associated to the activation of emotional long-term memory circuits.

Vectorial transfer of glucose across a membrane should be coupled with an energy dissipative potential [4, 98, 99, 100]. Glucose may signal for a more reversible promoter-induced form of genetic expression.

Oxidative phosphorylation contributes to the glucose-dependent phosphate uptake [101, 102, 103].

Circadian Clock in Mammals

In mammals, in addition to receiving signals from peripheral oscillators like the photosensitive retina, the hypothalamic and RHT (Retinohypothalamic Tract), the suprachiasmatic nucleus (SCN) also receives feedback from pineal gland to integrate circadian regulation of physiology and behavior.

Such peripheral oscillators can become uncoupled from the SCN if their specific needs dictate as occurs in liver, lung and skeletal muscle after entrainment by food [104].

The central pacemaker is in the SCN located in the ventral hypothalamus.

The neurons terminals at the hypothalamus suprachiasmatic nucleus have a biologic clock, synchronized by external signals like light and dark periods. The hypothalamus has neuronal sensors to monitor glucose fluctuations in blood.

The circadian clock controls adrenocorticotropic hormone (ACTH), cortisol/hydrocortisone, and melatonin. Sleep-related are prolactin, thyroid stimulating hormones (TSH) and growth hormone (GH). TSH and cortisol pulses are related to the sleeping superficial phases.

The release of ACTH is pulsatile with about 7-15 episodes per day. The stimulation of cortisol release occurs within 15 min of the surge in ACTH with follows a circadian rhythm sensitive to stress, light, sleep and disease.

In the early waking hours the release of cortisol increases but decline as the day progresses.

At the beginning of the night GH and prolactin show positively linked pulses during the first slow-wave of sleeping. Disorder of circadian patterns has been related to disease and the impaired endocrine function observed for aging.

Governed by the nervous system the hypothalamic neuro-hormones, through a capillary network, release at the anterior pituitary the tropic and target-organ hormones to stimulate a cyclic pattern, which maintain metabolic homeostasis.

The primary role of the cell-autonomous circadian clocks is to maintain their own 24-hour molecular rhythm and to drive the rhythmic expression of genes involved in physiology, metabolisms and behavior.

These systems provide internal temporal organization so that rhythmic metabolic activities became synchronized according to a circadian cycle. The latter by the entrainment process transmits signals like environmental light to the timekeeping apparatus to set the clock time to the specie diurnal-sleep cycle.

This match of local environment parameter also involves temperature compensation to synchronize metabolic rates between organs.

The daily light-dark cycle by secreting factors, TGF-α and Prokineticin, result in synchronizing rhythms of locomotor activity, behavior and metabolism, which act locally within the hypothalamus. This results in subservient oscillators in liver, kidney and peripheral tissues via humoral signals.

Two nuclear receptors, REV-ERBα and NURR1, connect circadian clock and mood regulation through dopaminergic neuronal action [105].

The expression levels of brain and muscle Arnt-like protein-1 (BMAL1; also known as MOP3 or Arnt3) are transcription factor known to regulate circadian rhythm are low at the beginning of a subjective night and over the day increase leading to formation of BMAL1-Clk heterodimers.

These heterodimers Per1, Per2 and Per3 drive the transcription of three mammalian period genes, which become phosphorylated by casein kinase-I-epsilon (CSNK-I-Epsilon), these forms are unstable and are degraded by ubiquitylation.

Evolution adjusted circadian pacemakers to geophysical time in order to adapt to obtain synchrony between increases in activity with the time of food availability and/or avoidance of predators.

Many problems in sleep patterns are related to an inability to properly entrain the circadian clock to the day-night cycle. Furthermore, disruptions of the sleep pattern due to jet lag or shift work are also related to the entrainment of the circadian clock [106]. Abnormalities in circadian behavior are referred as familial advanced sleep phase syndrome (FASPS) [107, 108].

Chapter 3

Molecular Details of Ligand Selectivity

The ionic radii: 0.072 nm for Mg^{2+}, 0.067 nm for Mn^{2+} and 0.074 nm for Zn^{2+} in the octahedral and 0.060 nm in tetrahedral coordination. These ions are smaller than 0.1 nm Ca^{2+}.

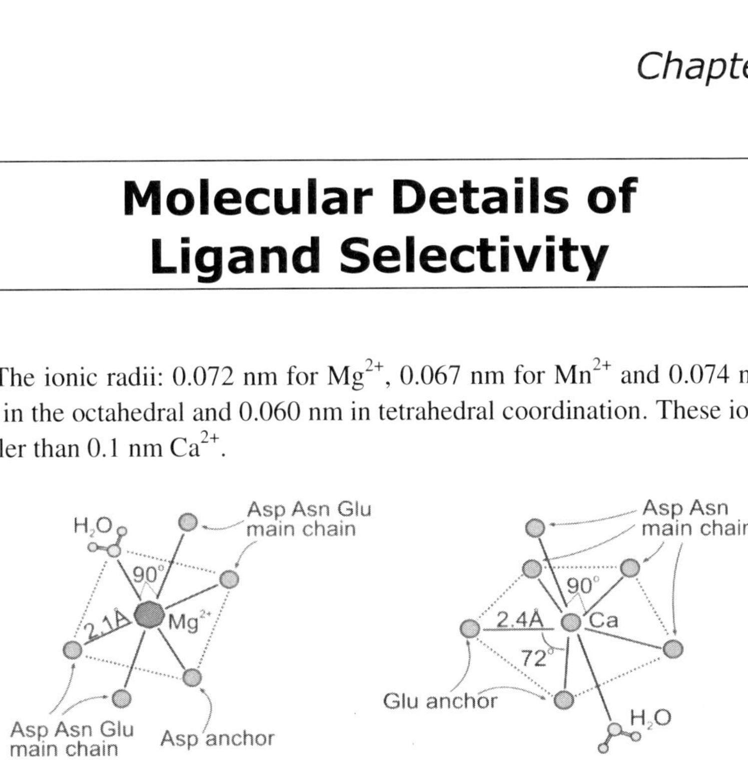

Figure 3. Molecular details of ligand selectivity. The location of the divalent metals on a simulated protein backbone was obtained by using a program. Aspartate (Asp) with the attachment of Mg^{2+} was use as the anchor position to calculate from this one the other ligand locations. Mg^{2+} was positioned with an apical 6th R-group to constitute an Asp-H_2O axis. At a distance of 0.21 nm between the divalent metal and each side chains: Asp, Asn, or Glu with H_2O resulted into an octahedral geometry with 5th and 6th coordinative position configuring the axis. Mn^{2+} also prefers octahedral ligand geometry, whereas Zn^{2+} has a preference for tetrahedral coordination. The obtained R-group configuration for a Ca^{2+}-binding site with 0.24 nm of distance configure a pentagonal-bipyramidal geometry between the divalent metal and each of the side-chains Asp or Asn (anchored to a Glu) and Glutamate (Glu) or Asp provides two atoms of oxygen for a bidentate ligand Ca binding into an EF-hand motif, troponin C (TnC), calmodulin (CaM), neuronal Ca^{2+} sensor proteins (NCS) and the S100 proteins.

The role of the Ion Ca^{2+} in Biological Signaling Controls

Ca^{2+} signaling functions to regulate many different cellular processes. It triggers on-set of life at the time of fertilization, controls development processes and once cells differentiated modulates many cellular processes, including cell death by larger normal increase of the ion either programed (apoptosis) or necrotic in stroke and cardiac ischemia. Ca^{2+} is low in resting cells, but a stimulus increases the ion concentration to reach a new phase of cellular activity. There are multiple variations on the Ca^{2+} concentration mechanism of function.

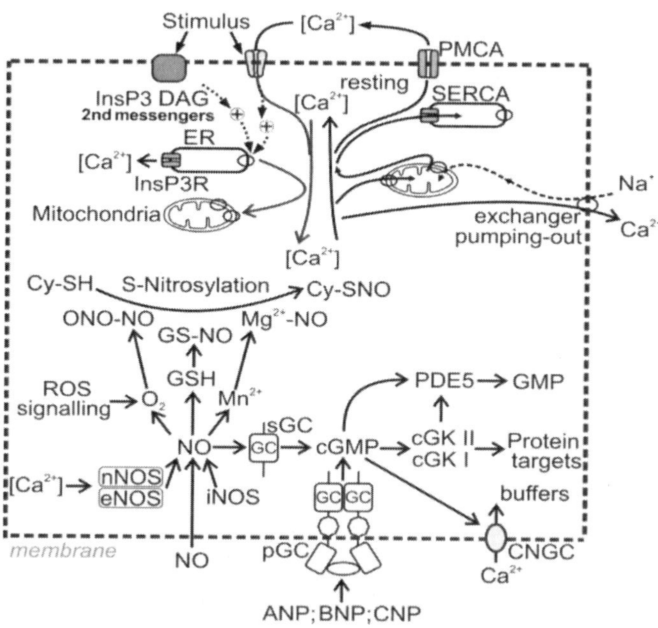

Figure 4. Ca^{2+} signaling mechanisms. A Nitric oxide (NO) paracrine signal could act by diffusion from other cells within the cell itself could be produced by different NO synthases (NOSs). Diacylglycerol (DAG), cyclic AMP, cyclic guanosine monophosphate (cGMP) and arachidonic acid acting from the cytoplasmic side open second messenger-operated channels (SMOCs) in the plasma membrane. Cell-surface receptors inositol 1,4,5-trisphosphate (InsP3), which then diffuses into the cell to activate the InsP3 receptor to release Ca^{2+} from the ER. The cells to reach the resting state activate the mitochondrial reuptake and the pumping of Ca^{2+} out of the cytoplasm. NO can stimulate soluble guanylyl cyclase to form the messenger cGMP signaling pathways to activate cyclic nucleotide-gated channels (CNGCs).

The cells cytoplasm maintain 100nM [Ca^{2+}], but after a stimulus the entrance of Ca^{2+} from channels or from the endogenous endoplasmic reticulum (ER) rises to 500nM the [Ca^{2+}]. The entering ions bind mostly into the cytoplasm buffers calbindin D-28k (CB) and parvalbumin (PV), and taken-up by mitochondria. The Ca^{2+} entry phase is terminated in a sequential manner by the Na^+/Ca^{2+}-exchanger, and the Ca^{2+} pumps: plasma membrane Ca^{2+}-ATPase (PMCA), sarco/endoplasmic reticulum Ca^{2+}-ATPase (SERCA) for the termination of the action spike.

The Ca^{2+} signaling include entry channels controlling the uptake from the outside and from internal stores, as well as the Ca^{2+} release to the outside. Buffers control operation range to prevent Ca^{2+} level leading to cell death. The Ca^{2+} pump and exchangers remove it from the cytoplasm to return in the internal stores. Signaling function require Ca^{2+} sensors and effectors.

Time domains: for exocytosis, μs; for contraction, ms; for metabolism, sec and for gene transcription, min.

Mn^{2+}, Zn^{2+} and Mg^{2+} could not activate CaM, the latter ion affects the specificity of CaM and its distribution among targets [109, 110].

Cyclic nucleotide-gated channels (CNGCs) activate either Ca^{2+} entrance or cyclic guanosine monophosphate (cGMP)-dependent protein kinase (cGK) and reactive oxygen species (ROS) signaling. cGMP is also formed by a plasma membrane guanylyl cyclase (pGC), which is part of the single membrane-spanning receptor activated by a variety of peptides. Ca^{2+}/calmodulin-dependent protein kinases (CaMKs), calcineurin, phosphorylase kinase, and myosin light chain kinase (MLCK), Ca^{2+}-promoted ras inactivator (CAPRI). Atrial natriuretic factor (ANP), C-type natriuretic factor (CNP) guanylin, brain natriuretic factor (BNP).

Calreticulin (CRT) is a low-affinity Ca^{2+}-binding protein that is located within the lumen of the endoplasmic reticulum (ER) and homologous proteins calnexin and calmegin have an acid C-terminal domain with a large number of low-affinity Ca^{2+}-binding sites. CRT terminates in the KDEL sequence, responsible for its binding retaining CRT into the ER lumen, binding capacity 20-30 mol of Ca^{2+}/mol of protein.

The buffer calbindin D-28k (CB) in enabled neurons for highly localized Ca^{2+} events which occur in spines, this ones have volume, which increase as a function of lack of Ca^{2+}-buffers. The spines increase the numbers of their synaptic connections and this neuronal miniaturization greatly enhances the signal processing capacity of the brain.

The calcium affinity of neural cell adhesion molecule (NCAM) has functional roles during nervous system development and in the adult after injury and in synaptic plasticity. It shows a closed-domain apo-like conformation, Isothermal titration calorimetry shows that two metal binding sites of N-CaM are occupied by Ca^{2+}, Mn^{2+} and Mg^{2+} at two metal binding sites, but the Kd(μM) Ca^{2+} 5.6 is half the Kd for Mn^{2+} that binds much more strongly than Mg^{2+} Kd=450 [111].

The Zn^{2+} ions do not bound to the Ca^{2+}-binding loops of N-CaM in a position normally occupied by Ca^{2+}.

Mg^{2+}/Ca^{2+} competition allows Mg^{2+} to switch-off CaM when Ca^{2+} decreases to resting levels, because the titration curve for Ca^{2+} shift to higher Ca^{2+} concentration and the extent of activation will be inversely related to the Mg^{2+} tendency to occupancy of N-CaM [112].

The Role of the Mg^{2+} in Biological Signaling Controls

Most of the cellular ATP is present as an MgATP complex, which is the substrate for many enzymes including ATPases, adenylyl cyclase (AC) and IRTK. Also Mg^{2+} regulates many enzymes and transporters, and the release of Ca^{2+} from the SR, modifying the activity of many ion channels in the plasma membrane.

The transcription factor DREAM (calsenilin/KChIP3), an EF-hand calcium-binding protein that represses transcription of prodynorphin and c-fos genes, when binding Mg^{2+} alters transcription.

Free Mg^{2+} is buffered at about 1mM cytosolic Mg^{2+}, leading to claims [112] that this ion does not play a direct signaling regulatory role.

Cytosolic free Mg^{2+} is regulated by plasma membrane and organelle transport, and by intracellular binding, is maintained far from electrochemical equilibrium.

β-adrenergic agonists stimulate a large Mg^{2+} efflux from perfused heart and isolated myocytes [113, 114]. Noradrenaline (NA)-induced Mg^{2+} efflux from the cell is blocked by removing extracellular Ca^{2+} [115].

The involved mechanism may depend of the alteration of Mg^{2+} buffering and the release of Mg^{2+} into the cytosol, which cannot exceed the maximum rate of efflux across the plasma membrane [116].

β-adrenergic agonists increase the NA responsiveness of adenylyl cyclase and the formed cyclic adenosine monophosphate (cAMP) induce from mitochondria the efflux of Mg^{2+} [2, 117, 118, 113], which is blocked by an inhibitor of the adenine nucleotide translocator.

It has been suggested that oxygenation increase the affinity for Mg^{2+} of Hb, allowing this protein to function as an Mg^{2+} carrier in the transition of oxyHb to deoxyHb. Cellular uptake of O_2 could deliver Mg^{2+} for cellular metabolic activation, at the synaptic cleft Mg^{2+} initiate responsiveness to presynaptic NA by activation of postsynaptic adenylyl cyclase [2]. The lack of important changes in cytosolic free Mg^{2+} could be explained by oxyHb deoxygenation providing for *on demand* of O_2 release with simultaneous delivery of Mg^{2+} at the cellular membrane and synaptic cleft *locus*. The high charge manifested by the poorly hydrated Mg^{2+}, when it is release from its complex state with a protein, ensures that its high activity even at low concentration could overcome the poor activity of the Mg^{2+} surrounded by a complete hydration shell.

Adrenergic responsiveness of α-receptor to NA or β-receptor to adrenaline depends on the release of Mg^{2+} by a change a decrease in the concentration of Mg^{2+} chelators ligands $Kd(MgATP)=[MgATP]/[ATP^{4-}][Mg^{2+}]=4x10^{-5}$ M a decrease in the Mg^{2+} chelators specie (or buffering system) increases the concentration of free Mg^{2+}. In the red cell deoxygenation results in a decrease of 2,3-DPG to form the complex 2,3-DPG deoxyHb releasing Mg^{2+}.

$Kd(Mg\text{-}2,3\text{-}DPG)=[Mg\text{-}2,3\text{-}DPG]/[Mg^{2+}][2,3\text{-}DPG]$ also a decrease in oxyHb to form deoxyHb. $Kd(Mg\text{-}oxyHb)=[Mg\text{-}oxyHb]/[Mg^{2+}][oxyHb]$.

Deoxygenation therefore releases Mg^{2+} from its 2,3-DPG and oxyHb complexes.

A large release of Mg^{2+} could occur in the brain by oxyHb conversion into deoxyHb specie. In brain, this effect is desirable to reduce the rate of glycolysis. In other tissue small relative changes in the Mg^{2+} ion released may not affect by much glycolysis and ion pumping [119].

An uncomplexed ligand or buffer available to Mg^{2+} could be influenced by Ca^{2+}, pH, and other ions that bind to the same sites as Mg^{2+}.

The free Mg^{2+} concentration was shown to change from 0.57mM to 1.9mM in between the change from oxygenated to deoxygenated cells, increasing the tendency from cells to release Mg^{2+} into the surrounding media [120].

Ca^{2+} could displace Mg^{2+} from mutual binding sites incrementing cytoplasmic Mg^{2+}.

Caffeine could produce a large decrease of calcium in sarcoplasmic reticulum (SR) increasing Mg^{2+} by 50% [121].

Nitric Oxide Synthase

The enzyme nitric oxide synthase (NOS) was first described in neurons (nNOS) is constitutively active, and regulation depends upon its induction by inflammatory receptors such as those that respond to interferon-γ (IFN-γ) or lipopolysaccharide (LPS). The glial cells nNOS bind to calmodulin (CaM) for activation. In the plasma membrane binds through its PDZ domains. In neurons, nNOS binds to the postsynaptic density (PSD) proteins such as PSD-93 and PSD-95.

The Role of Water

Changes in hydration shell of a protein occur because the tendency of the dipolar water molecules is to make hydrogen bonds with R-groups. The latter reactivity allows to chelate ions like Mg^{2+}. The tendency of a protein tor form complexes depend on the number and strength of partial charge of the R groups of the polypeptide chain available on the surface of the protein. Therefore a dynamics mechanism to form and break H-bonds allows proton-induced change in conformation of the protein that could relate to a ΔpH inducing ΔpKa, like these of R-histidines side chains.

The polarity of water molecules at room temperature and normal atmospheric pressure maintains an average n=3.4 water molecules H-bonded between themselves. Water has a larger calorific capacity than other liquids, due to its capacity to absorb heat by decreasing the number of the molecules of water, linked into an average cluster structure ($(H_2O)n<3.4$). The entropy of the aqueous system of increases as the more highly ordered arrays of water molecules are replaced by the less orderly poorly H-bonded in liquid water.

These H-bonds depleted water becomes uptake by the astrocyte to be eventually conducted outside of the neuron system itself. Hence, heat produced during ligand-induced changes of the protein conformation is retained by the released water.

The depleted H-bonded water when released outside of the body, at cooler temperatures follow the direction $(H_2O)n<3.4 \rightarrow (H_2O)_{vapor}$. The overall effect limits entropy increase, allowing to sustain the homeostatic state of temperature, which characterizes brain. The overall thermodynamic changes has to be plotted again the scale of the endergonic direction. The expected change depends of the number of molecules of water loss per protein molecule during the ligand-protein transition. To evaluate thermodynamic change in oxyHb-$(H_2O)n$, n=60 in transition to deoxyHb + $60H_2O$, has to be taken in account that even an small ΔH per molecule of water has to be multiplied by 60.

The transition between these differently ligand states of the protein allows for a sequential exergonic release of energy, decreasing equilibrium tendency at each of the ligand-protein complexes into a final equilibrium for predominance of the conformational changes.

The tendency of the water molecules to associate through hydrogen bonds is outweighed by the energetic push toward randomness. Differentiable ligand states of the same protein like oxyHb vs deoxyHb could complete a turnover cycle by thermodynamically balancing the exergonic with the endergonic senses.

The balancing could be provided by the change in the number of water molecules linked and release during transition from the R to T state of the protein conformation.

The breaking of hydrogen bonds between water molecules takes up energy from the system, requiring the input of energy from the surroundings.

Water molecules are not as highly oriented as those in cagelike shells around crystalline structures of nonpolar solutes and water, forming highly ordered cagelike clathrates. Hence, the input of energy to dissolve hydrophobic compounds in water produces by the ordering of water reduces entropy. Long chain fatty acids have hydrophobic alkyl chains surrounded by a shell of ordered water molecules.

The number of ordered water molecules, and therefore the magnitude of the entropy decrease, is proportional to the surface area of the hydrophobic solute enclosed within the cage of water molecules.

Figure shows that by shedding their hydration shells Mg^{2+} and Ca^{2+} could fit into pore-regions of transporter proteins. The mechanisms are similar. The energy released its larger for the smaller Mg^{2+}, which is surrounded by a more stable double hydration shell than Ca^{2+}.

A labile second layer is not usually involved in the mechanism of Ca^{2+} translocation.

The literature reports the calculated successive ΔE kcal/mol in $Mg(H_2O)n^{2+}$ by increasing the state of hydration by capturing of 1 molecule of H_2O and the Mg-O bond distances [98].

a) $[Mg(H_2O)_4](H_2O)^{2+}$: Mg-O=0.194 nm, ΔE=23.7. b) $[Mg(H_2O)_6](H_2O)^{2+}$: Mg-O=0.204 nm, ΔE=18, c) $[Mg(H_2O)_6](H_2O)_{12}^{2+}$: Mg-O=0.207 nm, ΔE=13.1.

Similarly the values for $Ca(H_2O)n^{2+}$ are a) $[Ca(H_2O)_4](H_2O)^{2+}$: Ca-O= 2.25nm, ΔE=20.7 b) $[Ca(H_2O)_6](H_2O)^{2+}$: Ca-O= 2.32nm, ΔE=16.3, c) $[Ca(H_2O)_6](H_2O)_{12}{}^{2+}$: Ca-O= 2.35nm, ΔE=12.9.

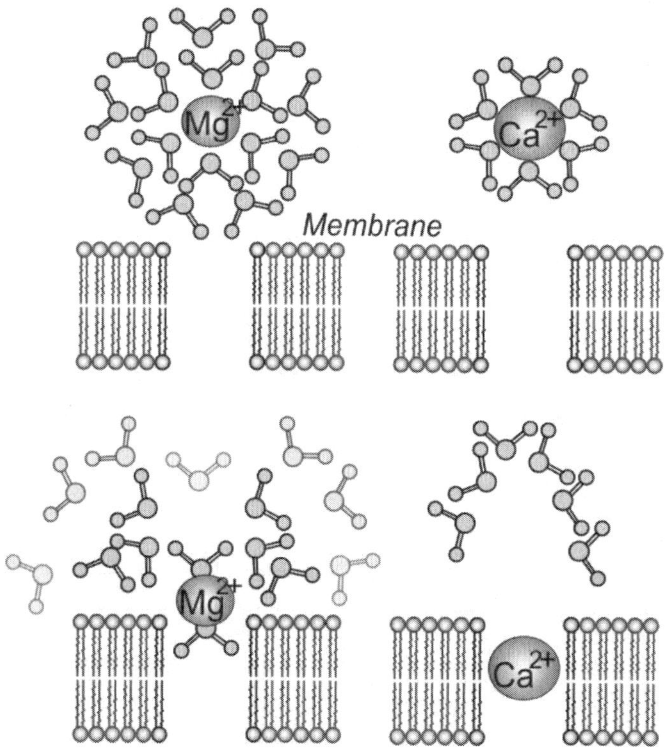

Figure 5. Illustration of the layers within the hydration shells of Mg^{2+} and Ca^{2+}. Ca^{2+} is shown with only one layer because the 2nd one usually does not participate in the fitting within a channel. These divalent metals have a tendency to coordinate two or more residual side groups of a polypeptide chain. The strength of these chelate links maintains the quaternary structure of protein.

Total energy involved in a transition could be inferred from the gaining or loss of water molecules at the level of the conformational changes of the protein within a complex.

Total water binding energies (E_{tot}) [kcal/mol] could be obtained by summing-up the number of H_2O molecules uptake from the hydration shell of the divalent metal.

The loss of H_2O by the hydration shell of the divalent metal could become the exergonic sense because is supported by coupling with uptake of water molecules by the protein. The mechanism is supported by a decrease in the number of molecules of H_2O aggregated into a cluster, which would have the tendency to become progressively depleted. Cerebrospinal fluid (CSF) turnover could replace smaller water aggregates for larger ones.

The cellular Mg^{2+} vs Ca^{2+} ionic concentration of 1000/1 ratios leads to the assumption that the role of Mg^{2+} in signaling involves an incognita. This because small changes in Ca^{2+} levels could be significant for *signaling* whereas the large Mg^{2+} concentration prevented such a role.

Conformational changes of structure-function related to hydration state are not only functional to the Hb-plasticity at the quaternary structure level.

The Cα rotation promoted by the loss of Mg^{2+} facilitates the side chains His (histidine) 143 and His 92 of Hb to conform a binding site for 2,3-DPG, when moving away and collapsing it to allow Mg^{2+} binding for His 92 to move and reach the interacting distance with the iron Heme.

Evolution has also developed full compartmentalization separating the hydration from the dehydration site as seen in the structure of the thermotoga maritima cellobiose-binding protein (tmCBP). The binding site is bipartite a solvent-excluded region (sub-site one), which is close to a solvent-filled cavity (sub-site two) for specific and semi-specific ligand recognition. The rich H-bonding potential of water molecules may facilitate the ordering and the plasticity of this highly conserved periplasmic binding protein (PBP) [122].

The Relationship of Protein Conformational Changes to Mechanisms of Energy Transduction

Studies on the uncoupling mechanism of 2,4-dinitro phenol (DNP) on oxidate phosphorylation coupled to yeast respiration [101], resulted in the loss of ATP-synthase activity, but the ATPase was isolated using 2,4-DNP as a marker. It was unmasked a particle bound DNP-stimulated Mg^{2+}-activated ATPase, which showed preservation of affinity for ADP, but inactive as ATP-synthase [103]. A particulate preparation of purified spinach chloroplast showed a light-dependent ADP-phosphorylation, but in apparent violation of microscopic reversibility did not show any detectable reversibility /ATPase [102].

Incubation with Ca^{2+} uncouples the particles, which manifested a light-dependent ATPase. Incubation with Mg^{2+} plus cysteine induced a light-triggered ATPase. EDTA-treatment of chloroplast particles and subsequent centrifugation resulted into a 2 fraction resolution, a residue and its supernatant denominated chloroplast coupling factor one (CF_1), both individually light-inactive. Both fractions were mixed and, by a time-dependent incubation, resulted in recombined particles. These ones recovered a large percent of the 3 differently light-coupling states of the initial chloroplast particles.

CF_1 also denominated CF1-ATPase was purified. It was shown a divalent metal dependence for coupling to the photo active electron transport chain, present in the resolution-residue. Therefore, the CF_1-protein could couple with electron transport to reach a high energy intermediate state, capable to drive ADP phosphorylation. The relationship between changes in high energy and conformational state was shown capable to modify affinity contants at the active site of CF_1-ATPase.

Each of the three states of coupling was characterized for their differentiable affinity for ADP and GDP. The latter acted as competitive inhibitors of the light-dependent and the light-triggered ATPases. Each of these specific incubations also allowed characterizing the residual reaction of photophosphorylation. In these, the affinity constants of CF_1-ATPase for ADP and GDP, as substrates of the phosphorylation reaction, were shown to be the same as when acting as inhibitors of each of the two ATPases.

At that time was prevalent the assumption that in both: chloroplast and mitochondria, the electron-transfer was linked by a covalent bond to phosphorylation. The preceeding results turned-down the previous dominant hypothesys that a covalent bond transferring was involved in the formation of the high energy intermediate. Hence, the importance of results predicting that high energy-transfer was mediated by the conformational change in CF_1. The induce-fit mechanisms operating in coupling were shown driven by coordinative rather than covalent chemistry. Consequently, was predicted that coordinative-mediated high-energy intermediates allowed energy transduction with ATP synthesis. Also, which the transduction use of the energy ATP hydrolysis was capable to operate and mediate electrogenic membrane processes [83].

The lack of detectable reversibility at the active site of the ATP synthase-ATPase level of fully couple photophosphorylation indicated a non-equilibrium energy transduction. The open system nature of these processes allows the input and output of energy and molecules to support dissipative energy potentials.

Prigogine's evaluation of biological dissipative structures did not include biological models. The characterization of CF1-ATPase indicated that coupling was mediated by delocalization of activation energy (E_a). The latter, required driving a vectorial sequence of transition states, between two conformational forms of the protein. Prigogine proposed that a system with a minimum entropy production, could steadly go down to a minimum were it stays, supporting for a time a steady state. This non-linear probable state refers to perturbations of thermodynamic forces, too small to in-validate a principle of maximum entropy production. Hence, the need to discern for vectorial systems a mechanism, which could prevents microscopic reversibility. The latter, has to be overpassed in order to prevent the entangling of the process directionality. To solve the latter, lead to analyze the change of oxy to deoxy in Hb, in function of its multiple-equilibrium, including the turnover-dependent changes in its hydration shell. This could be driven by depleting the H-bonds present in the surrounding water cluster ($(H_2O)n$), and functions to expel entropy-out in an open system. Therefore, coupling by the conformational change in a protein, leads to a discontinuos path in which exists differentiation of space-time between a ligand uptake process (a revolving door) and that of its release.

Membrane bound ATPases by turning over an axis, show the revolving door, which could preserve in an open system, an outside supported dissipative potential. Provided that the coupling is available for H-bond depleted water exclusion, which prevents entropy accumulation, which otherwise could drive microscopic reversibility.

A model, in which the energy of an electrochemical gradient of proton ions would not be dissipated, implies a change in H^+ dissociation constants (ΔpKa) [83] during the conformational change of a protein.

This should involve the side chains of the amino acid sequence of a polypeptide chain (R-groups), during turnover of two or more stages of alternative coupling, allowing transition events a high energy role.

This would be the case if the energy associated with the release of a hydrogen ion on the course of ATP hydrolysis, which instead of being dissipated immediately as heat of neutralization, could be conserved.

Accordingly, it was found that $AT^{32}P$ could be formed by the reversal of the normal sequence of light-dependent and light-triggered ATPases of chloroplast. Also the Na^+ and K^+-stimulated ATPase [123] could form ATP mediating reversal into the backward direction of the active ion transport in red cells [124].

The modification of the quaternary structure of protein complexes by turnover between two or more ligands, reveal energy dissipative events. This could be coupled to the breaking and reconstitution of H-bonds, changing the number of water within the hydration shell [125, 126]. Supporting the same, the number of H-bonds could decrease within the surrounding water cluster.

Chapter 4

Dynamics of Ligands Exchanges during Hb Deoxygenation

An increased order of H_2O molecules in a solvation layer allows a decrease in the entropy of the water. If nonpolar groups cluster together, the extent of the solvation layer decreases, with changes in entropic state within the molecule, since the involved groups lack surface exposure. The hydrophobic side chains of amino acid tend to cluster in a protein interior, and the H-bonds tend to be formed by the same entropic effect.

A solvation layer also forms around polar groups, in which H-bonds per unit mass are greater for proteins in pure water than for any other *solvent*. The release H_2O molecules from the structured water, provides an entropic driving force for folding. OxyHb in R-form loss of 60 molecules of H_2O participates in the deoxygenation intramolecular interactions, which results in the T-form conformation of deoxyHb.

Most of the net change in the internal free energy of a protein, participating in a conformational change is related to weak interactions like Δ H-bonds, charge attractions, ΔpKa, etc. Considering all forces, the conformational change is mainly drive by the increased entropy associated to the hydrophobic tendency of the protein, to regain a tighter folding. This counterbalances the large energy requirement for conformational-associated loss of ligand water from within the protein itself.

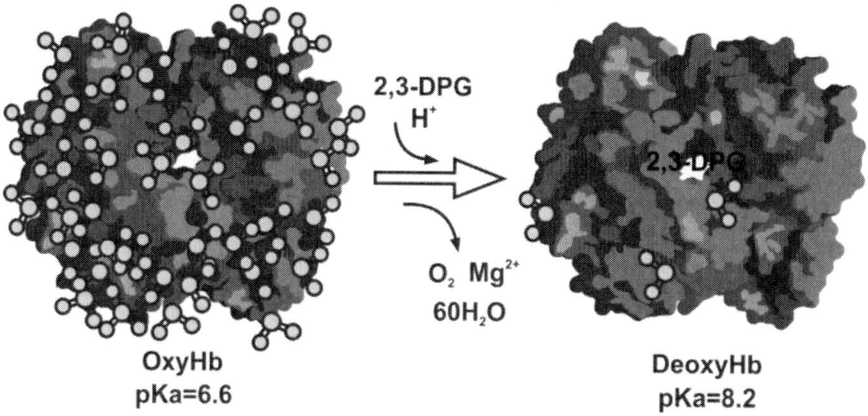

Figure 6. The hydration shell of Hb is shown releasing water. The event is driven by transition from the R to T-form of Hb, during its exchange of the ligands O_2 and Mg^{2+} in oxyHb for 2,3-DPG in deoxyHb. The x-ray diffraction pattern of the crystal of oxyHb shows 60 ligand molecules of H_2O, firmly bound as if were parts of the structure.

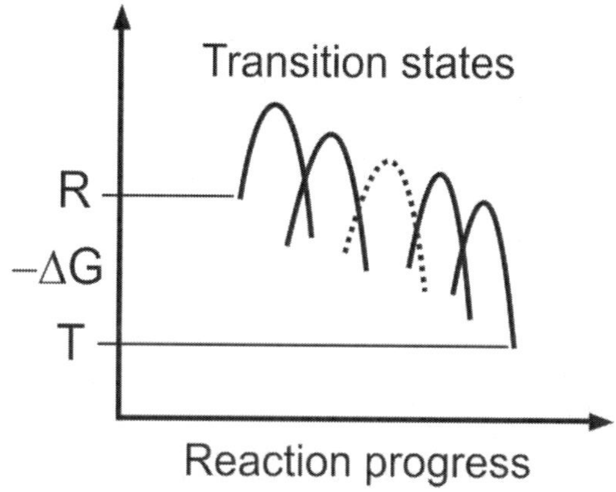

Figure 7. Integration of ligand-exchanges during Hb deoxygenation. Solid lines peaks represent energy coupled transition states (TS), integrated by separated contributions to the overall energy of activation (*E*a). The broken line has been used to represent a short-lived transition state, which makes possible to disconnect the initial steps from the final ones. The broken line peak represents a differentiable event conditioning reaction progress to a vectorial kinetic.

Crystals of RNA, DNA and proteins have tightly bound water molecules, which are not osmotically active, as is these were part of the structure itself, but with different properties than the bulk water.

Transition from the R to the T-form changes H_2O association constant of Hb $\Delta pKa(H_2O)$. Water molecules can also be detected as ligands of a protein in solution by nuclear magnetic resonance, and/or water-glycerol equilibrium studies. It is proposed that the change in a protein hydration shell is essential, also for many enzymes in the cycle enzyme-substrate transition to enzyme-product.

Figure 7 illustrates a TS sequence of partial conformational states separated by progression of a multiple-equilibrium to reach full conformational change, involving the overall *E*a.

Conformational changes are driven by the formation of several differentiable ligand-protein complexes. A thermodynamic open-system allows that this pattern of reaction to progress without significant reversibility from Hb oxygenation in lungs to deoxygenation in tissues.

The thermodynamic change by the breaking and re-arrangement of H-bonds within the molecule is not subject to thermic tendency conditioning molecular microscopic behaviour and therefore could function in terms of vectorial kinetic.

Figure 7 illustrates that the exergonic change has to dominate in both direction sequences of Hb interconversion, T to R (in the lung), or its reverse R to T (in tissues). The energy potential reverses as a function of higher pO_2 and pH, at the lungs decreasing at the tissue level. The multiple equilibrium in deoxygenation involves dissociation of the ligands 4 O_2, 2 to $4Mg^{2+}$ and $60H_2O$, and association of 2,3-DPG, and H^+. The pKa of the R-form (oxyHb) changes from 6.6 to 8.2 in the T-form (deoxyHb).

The H_2O became released as single molecules and, therefore, tend to mix with water cluster decreasing its H-bond content. The water expelled-out from the body drives heat-dissipation, and therefore becomes a carrier of increasing entropy into surrounding cluster-H_2O. This coupling changes the Hb system tendencies by H-bond number thermodynamic balancing of oxygenation vs deoxygenation [125, 126]. Water binding in the direction of facilitating the R state is exergonic. Release of water should be endergonic, but overall there are exergonic contribution of H^+ and 2,3-DPG binding. The tendency of the greater effective charge of Mg^{2+} allows its ligand to protein state to be broken to be released.

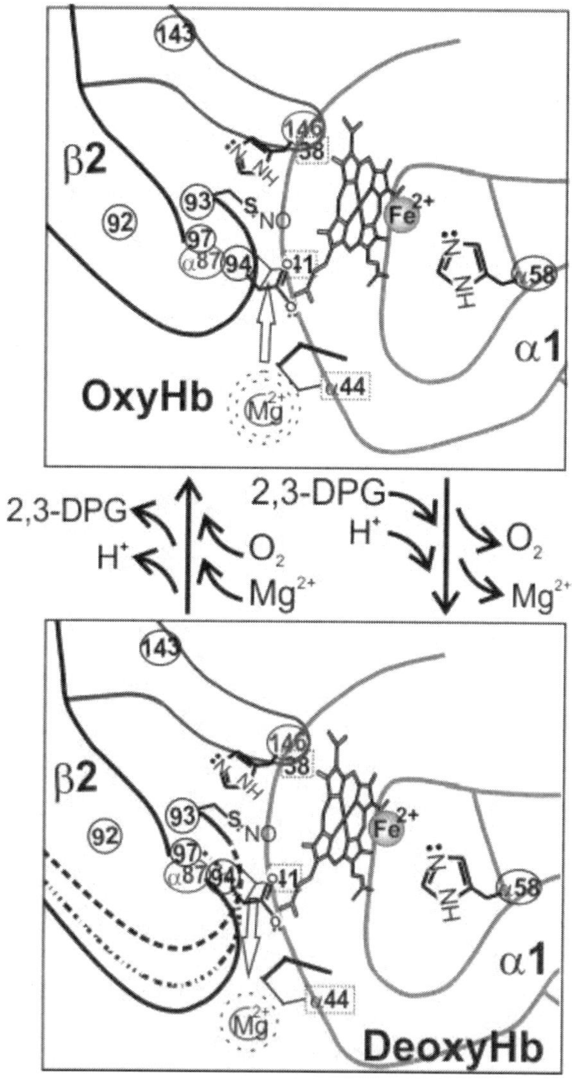

Figure 8. Structural turnover depending of the groups proline α_1/α_2 44 for control of Mg^{2+} access to chelating-hydrophilic region. Figure illustrates that at the lungs the deoxyHb conformational change to oxyHb involves the sliding of $\alpha_2\beta_2$ vs $\beta_1\alpha_1$, which allow entrance of two to four Mg^{2+} ions fully hydrated into the hydrophilic region. Chelation of Mg^{2+} occurs with the loss of most of its hydration shell and the Hb increases its hydration shell by about 60 H_2O molecules. The sliding displacement of R-groups allows for endogenous restructuration of coordinative bonds, overcoming thermic isotropic tendencies.

The quaternary structure of the protein is trapped in a latent/entropic state, in which side chains/R-groups move in random translational, rotational, and vibrational kinetics equivalent to 0.7kcal/mol at 37°C.

The latent/thermal-state could be overcome by coupling with water molecular dynamics, to activate functional relationships for conferral vectorial tendencies to side chain reactivity [8, 82, 83, 127].

At the brain, oxyHb not only functions transporting O_2 for release in the synapses, but does the same for Mg^{2+}. The conformational change associated to the quaternary structure transition of oxyHb from a relax (R) to a tense (T) form in deoxyHb, involves the quaternary structure of the protein to release 60 H_2O molecules. Thus, facilitating that Mg^{2+} would be released with a decrease number of H_2O molecules in its hydration shell. Effect facilitated by the increase of deoxyHb pKa=8.2 to oxyHb pKa=6.6. The tendency to decrease H^+ dissociation reduces the tendency to bind Mg^{2+}, which lacks water in its remaining hydration shell, pulls-out 60 water molecules from the protein hydration shell.

The conformational change decreasing affinity for O_2 leads the chelated R-His groups to release Mg^{2+}. The mechanism involves an amphiprotic response by the displacing of Mg^{2+} through the increasing affinity for H^+ changing pKa to 8.2, during the imidazole ring restructuring to imidazolium ($-N:- + H^+ \rightarrow =NH^{(+)}-$).

Hence, differentiating this *nascent* Mg^{2+}, denomination suggested by similitude with *nascent* chloride. The nascent Mg^{2+} could be differentiated from total Mg^{2+} by the smaller number of H_2O coordinated to the ion, reducing its overall size. This increases the *effective charge* of the Mg^{2+} ion released from its chelating interaction with the negative charged groups of the protein.

The sliding of the $\beta_2\alpha_2$ vs $\beta_1\alpha_1$ allows that proline α_1/α_2 44 Hb by not having free $C\alpha$ rotation could be displaced to block access to the hydrophilic crevices within Hb molecular interior.

Figure 8 illustrates the sliding of the $\beta_2\alpha_2$ vs $\beta_1\alpha_1$ in the transition of T to R that displaces symmetrically the two Pro α 44 elongated distances to β 94. This opens the access of fully hydrated Mg^{2+} (enclosed within a second hydration shell) to the hydrophilic crevices.

In the R to T sense an Mg^{2+} sliding allows closing the distance between Pro α 44 and the β chains, preventing that a full hydrated Mg^{2+} could enter, but still allowing the nascent Mg^{2+} to leave the crevices in the Hb molecule.

The interconversion of oxy to deoxy Hb creates, by the sliding of $\alpha_2\beta_2$ vs $\beta_1\alpha_1$ chains a modification of the proline α_1/α_2 44 salt links in the path for ions, like Mg^{2+}, to reach the hydrophilic multiple histidine configurated space (fig. 8).

After mutual inclusion of O_2 and Mg^{2+} stabilization of the conformational change into the R (relax) form $[(O_2)_4Hb(Mg)_2].(H_2O)_R$ reverse the sliding and in oxyHb two proline α 44 moves to close both gaps. This conformational change represents the less-reversible thermodynamic event for each sense of Hb turnover.

Dynamics of the Quaternary Structure of Hb during Oxygenation vs Deoxygenation

A ligand to one site of an allosteric protein affects the binding of other ligands at different sites. When a normal ligand and the modulator are identical, the interaction is denominated homotropic for other molecule than the normal ligand the interaction is heterotropic for some proteins with two or more modulators, both forms of interactions are present.

Cooperative conformational changes depend on variations in the structural stability of different parts of a protein. The allosteric protein may usually have stable segments in proximity with those capable of frequent changes in conformation. A moving part of the protein's binding site may be stabilized by a ligand binding [128, 129], and lead to a change in the configuration of bound water affecting the structure and activity proteins [130, 131].

Heterotropic allosteric modulation was shown for 2,3-diphosphoglycerate (2,3-DPG) [132]. 2,3-DPG cross-links subunits β_1 with β_2 of deoxyHb, at the tissues low pH a central cavity of the Hb tetramer became structured as an binding site for 2,3-DPG [133]. The latter, became surrounded by the terminal amino acid, Lys 82, and the protonated imidazole rings of histidines His 143 and His 2, in the β_1 and β_2 chains, which bear two NH bonds and show overall positive charge.

The Amphoteric Response of Histidine Side Chains

Imidazole is a highly polar compound as evidenced by a calculated dipole of 3.61D (debye), which is highly soluble in water. The compound is classified as aromatic due to the presence of a sextet of π-electrons, consisting of a pair of electrons from the protonated nitrogen atom and one from each of the remaining four atoms of the ring. Imidazole has resonance forms that facilitate amphoteric functions.

A single rigid plane with the oxygen atom of the carbonyl group trans-position to the H atom of the amide nitrogen configures the peptide group, in which the partial double-bond of C-N bonds could not rotate freely. This allows depicting consecutive planes sharing in between a common point of rotation at the N-Cα and the Cα-C bonds.

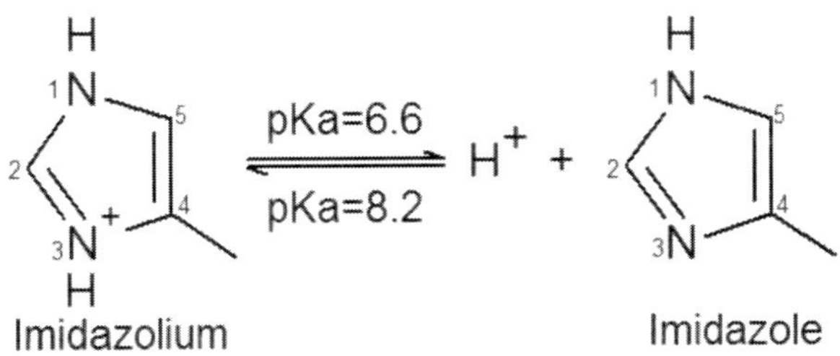

Figure 9. Imidazolium-imidazole transition. DeoxyHb pKa=8.2 for interconversion to oxyHb pKa=6.6, the histidine (His) side groups dissociate an H^+ from their ring to became imidazole R-groups, characterizing the relax form (R) of the protein quaternary structure oxyHb. The deprotonated imidazole is an anion pKa=7 with the dissociable H^+ became located on N-1, when two His-R-groups are proximal could manifest coordinative bonds with a divalent metal like Mg^{2+} to form a chelate.

As a base by protonation became the symmetric imidazolium cation with the basic site at N-3 pK_{BH}^+ (corresponds to the pK_a of a conjugate acid).

The R-groups participants in H^+ release or Bohr effect have been widly reported and allows to molecular characterization of proton disociated Hb as the R configuration within oxyHb. Interconversion of deoxyHb pKa=8.2 to oxyHb pKa=6.6 releases Bohr protons from R-groups within the two β chains: from Cys (cysteine) β_1 and $\beta_2$93, plus that from Asp (aspartate) β_1 and $\beta_2$94 [134, 135].

OxyHb is the acid form of Hb forming the Mg-Hb complex. At the brain, pH homeostasis prevents a H^+ driving transition but an increases in the 2,3-DPG concentration in the erythrocyte favors the change to deoxyHb.

The Hb-transition involves that the imidazole rings tend to attract H^+ to become imidazolium rings with decreased tendency to coordinate to the divalent metal. Hence, the oxyHb-Mg complex could be driven by 2,3-DPG binding to form the protonated deoxyHb-2,3-DPG complex, releasing free Mg^{2+}.

The mutually inclusive interactions of divalent metals and pO_2 allowing simultaneous exclusion of 2,3-DPG and H^+. According to saturation ranges of free (dissociated) divalent metals (Zn^{2+} or Mg^{2+}) could induce a coordinative-fit of ligand R-groups of Hb conforming two chelating sites.

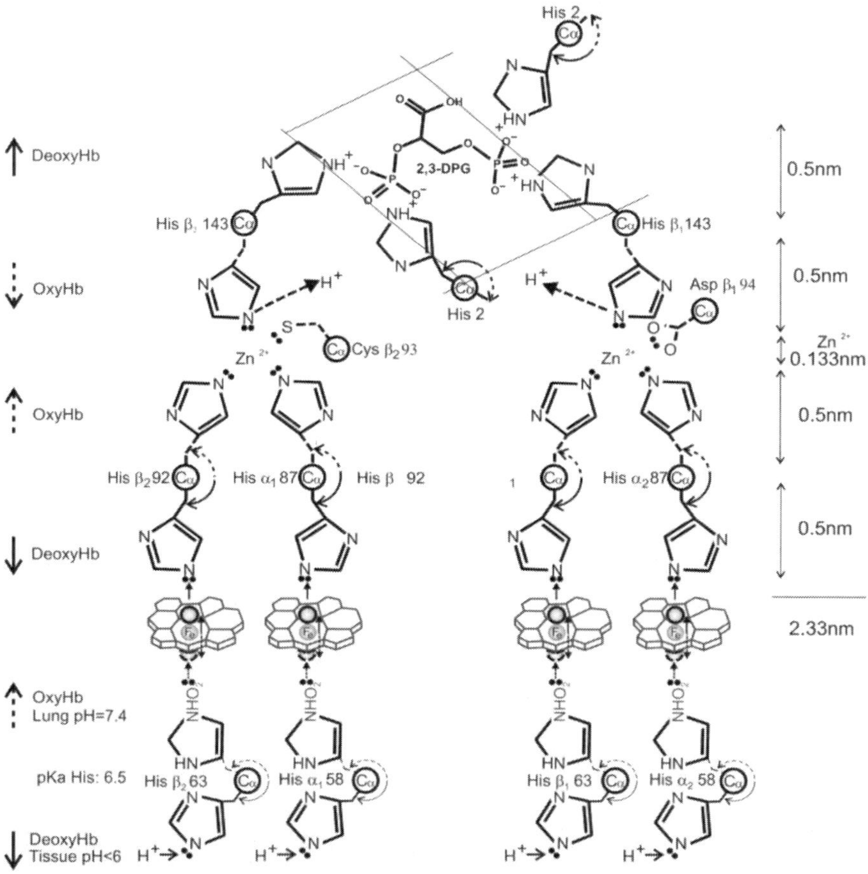

Figure 10. Dynamics of Mg^{2+}-chelation by His R-groups at the interface of β_1 α_2 chains of Hb. A second chelating site at β_2 α_1 chains is symmetric. Histidine (His) side chains extend 0.5nm and their Cα turn-around covered a distance of 1nm. For either His $\beta_2$143 or His $\beta_1$143 in oxyHb pKa=6.2 allow the imidazole form to manifest a negative charge capable to exercise attraction for the Fe^{2+}. The Fe^{2+} atoms move from their position at the Heme planes in the deoxyHb state (●) to allow their reactive position for oxygenation (◌). The oxygenation sequence (shown in broken lines: - - - -). Deoxygenation pathway (shown in solid lines: —) the R-groups shift-out of the chelation complex.

The oxygenation sequence (shown in broken lines: - - - -) predicts a physiological cooperative number n=2.8 for an oxygen saturation curve of Hb. This cooperativity number sums up the increase in affinity for O_2 and divalent metals Me^{2+} to a decrease in affinity for 2,3 DPG and H^+. Deoxygenation pathway (shown in solid lines: ——) the R-groups shift-out of the chelation complex.

A space to coordinate O_2 in between His $\beta_1$63 or His $\alpha_2$58, and their respective Fe^{2+}-Heme is stabilized. Deoxygenation deoxyHb pKa=8.2 results in the imidazole capture of a H^+ to became the positive charged imidazolium form to became attracted and rotate to reach and attract the negatively charged 2,3-DPG into induce-fit configuration of an antagonistic site to O_2 binding.

At tissue low pH, the tendency of divalent metals to became surrounded by hydration spheres, became an exergonic contribution to their release as ligand from Hb. Effect, facilitating that β_2 and β_1 His 143 capture an H^+ to became positively charged and by rotating along their Cα to participate in the conformation of a binding site for 2,3-DPG. Along the blood path to the tissues, progressive accumulation of 2,3-DPG would provide a constant decrease in the ratio of oxy- to deoxyHb.

Oxygenation requires the loss of the 2,3-DPG binding site, at the pH of lungs the protons dissociate from R-histidines imidazolium positive to manifest the reactivity of the negatively charged nitrogen atoms at both His 143 [136], which now negatively charged and similarly to the His β_1 and β_2 leave from the collapsing site unable to bind to 2,3-DPG.

Figure 10 shows that Cα rotation allows to couple the endergonic dissociation of 2,3-DPG with the exergonic association of Mg^{2+} or Zn^{2+}. Oxygenation in the red cell promote Mg^{2+} coordination to Cys $\beta_2$93 inducing fit of His $\beta_2$92, and facilitate the movement of the F-helix toward the H-helix [99, 137, 138, 139]. Thermodynamics considerations indicate that modeling of Hb has to balance oxygenation vs. deoxygenation for an energy efficiency coupling between the dissociation of 2,3-DPG from deoxyHb ΔH=5.7kcal/mol and the binding of Zn^{2+} with a ΔH=9.7kcal/mol per 2 Zn^{2+} atoms [140]. The solvation or hydration interaction between a solute like Hb and the solvent media of the cytoplasm allows an Hb hydration shell. The tendency of a divalent metal to acquire a hydration sphere could facilitate Me^{2+} and O_2 dissociation from oxy to deoxy. The model graphics patterns a solution involving the symmetric turn-around of both His β 92 and both His α 87, for dynamic transfer of the structural changes expanding the 2nm required for Heme to Heme interaction.

A Modified Woolf-Augustinsson-Hofstee (WAH) Plot

%saturation on the ordinate versus %saturation/$(pO_2)^{2.3}$ on abscissa, figure allowed to obtain an straight lines intersection at the theoretical 100% saturation on the ordinate.

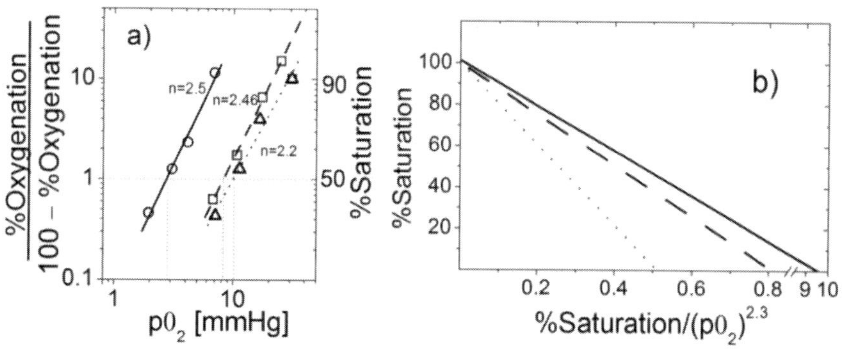

Figure 11 a) b). A modified Woolf-Augustinsson-Hofstee (WAH) plot. Graphic of the 2,3-DPG effect on the %-Oxygenation of stripped Hb. Adapted data [141] for the control: absence of 2,3-DPG (——); addition: +4×10mM of 2,3-DPG (- - - -) and +6×10mM of 2,3-DPG (····) were plotted according to: a) to Hill equation and b) a modified Woolf-Augustinsson-Hofstee plot in where the abscissa shows pO_2 at the exponential 2.3, value obtained from the Hill Plot. Introduction of the exponential is a solution to include cooperativity in order to transform curves into straight lines. Control values: (——) P_{50}=2.8mmHg, cooperativity n=2.5; in presence of +4×10mM of 2,3-DPG; (- - - -) values: P_{50}=8mmHg, cooperativity n=2.46 and +6×10mM of 2,3-DPG (····) P_{50}=10mmHg cooperativity n=2.2.

Treatment to obtain human Hb stripped-out of 2,3-DPG and other ligands increases Hb affinity for O_2, decreasing $P_{50}(O_2)$ 2.8mmHg [141]. The figure 11 illustrates a Hill plot showing that addition of 2,3-DPG [95] decreases Hb affinity for O_2 by increasing P_{50}, but without affecting Hb cooperativity for O_2.

The plot indicates that under the ideal condition of increasing pO_2 until reaching a theoretical maximum O_2-saturation, the presence of a constant concentration of 2,3-DPG may not prevent the homotropic Hb response of forming oxyHb.

A mutual exclusion between pO_2 and 2,3-DPG could lead to predict that a theoretical increase in pO_2, could reduce an inhibitory effect of a constant 2,3-DPG concentration [95].

Under physiological conditions Hb binds 2,3-DPG into a specific site and forms a complex in which the Hb quaternary structure, became stabilized as the deoxyHb form by the release of O_2.

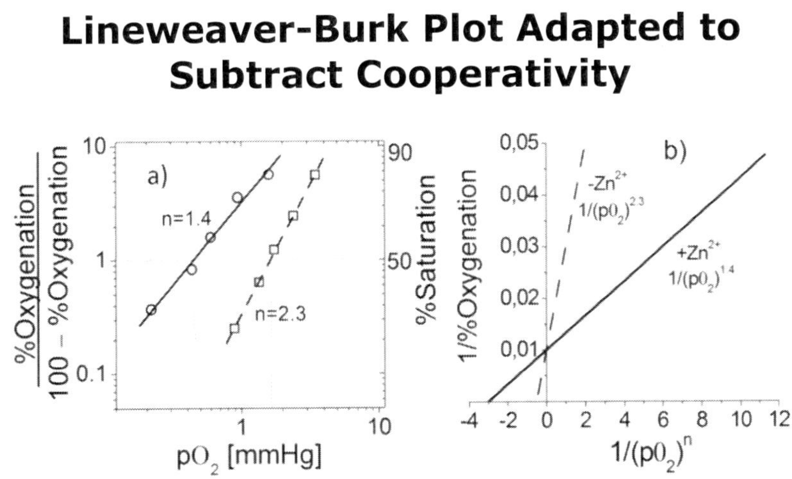

Figure 12. Graphics of a lineweaver-Burk plot of the effect of Zn^{2+} on the %-Oxygenation of stripped Hb. Adapted data for the control: absence of Zn^{2+} (- - - -), values: P_{50}=1.6mmHg; and addition of 4.47×10^{-2} mM Zn^{2+} (——), values: P_{50}=0.5mmHg was plotted according to a) Hill equation and b) a modified doble reciprocal Lineweaver-Burk plot in were the abscissa shows pO2 at the exponentials values obtained for the individual curves plotted according to Hill. Thus, the $-Zn^{2+}$ vs $1/(pO_2)^{2.3}$ and $+Zn^{2+}$ vs $1/(pO_2)^{1.4}$. The exponentials on pressure of O_2 correspond to the conversion factor required to obtain from sigmoidal data the plotted straight lines.

A double reciprocal plot Lineweaver-Burk was adapted to subtract cooperativity by introducing the cooperativity n-value for each curve in the abscissa $1/(pO_2)^n$. The plot becomes 1/%Oxygenation versus $1/(pO_2)^n$ and allowed to obtain intersection of straight lines for a theoretical maximal $1/(pO_2)^{1.5}$ and $1/(pO_2)^{2.3}$. Hence, at a theoretical maximal pO_2 ($1/pO_2=1/\infty\rightarrow0$) saturation of Hb by O_2 would be reached even in the absence of Zn^{2+}. Hence, the activatory effect was to decrease the control value $S_{0.5}$=1.64mmHg by addition of Zn^{2+} to $S_{0.5}$=0.5mmHg [141]. The strip method to free of ligands the Hb molecule appears to reduce steric hindrance. Even so Zn^{2+} becomes a strong enhancer of striped Hb affinity for O_2. The effects of 2,3-DPG decreasing affinity and Zn^{2+} increasing affinity of Hb for O_2, suggest by allosteric considerations alone that a single site for both actions, should be discarded. However, even if solving by classical competitive interaction could not be applied, other treatments were tested. The nullification of 2,3-DPG inhibitory and of the activatory Zn^{2+} actions, became evident when the plotting allow to assume a tendency for pO_2 to reach the equivalent of infinite concentration. The figure illustrates that because these ligands were at an invariant concentration, could be displaced from Hb by a competitive mechanism involving pO_2 increment.

The 2,3-DPG-Hb complex of much lower affinity for O_2 than isolated Hb could be regarded as the physiological T structure [142, 143]. Deoxygenation increase the concentration of free Mg^{2+} ion showing that 2,3-DPG and Mg^{2+} were mutually excluding ligand of Hb. Mg^{2+} and O_2 were mutually inclusive by each one increasing affinity for the other [120].

Competition between O_2 with inhibitor 2,3-DPG do not occur at the same site but may result from this heterotropic ligand tendency to bind and stabilize one or more R-groups restricting O_2-saturation of the Heme site of Hb. Accordingly, increasing pO_2 would eventually release the hindrance preventing O_2-binding to the Heme. The activatory effect by Zn^{2+} implicates that the effect of divalent metal is to decrease the concentration of O_2 required to remove the hindrance effect. Hence, a model was developing that shows that O_2 *per se* has full homotropic capability for the interconversion of deoxy- to oxyHb. The competition for a modulatory site, which control hindrance, could explain Hb heterotropic reactivity with ligands decreasing or increasing affinity for O_2. These dynamics imply that the modulatory sites could not be fully covalently structured, but are rather structures supported by coordinative bonding. The affinity of Hb for the ligands Mg^{2+} and/or Zn^{2+} supports the tendency to conform two coordinative structures of Hb, which fit the theoretical and physiological findings. The plasticity of this divalent metal or sugar phosphates modulated sites becomes evident in term of a glucose sensor role for the Hb-erythrocyte system [144].

Glucose as a sensor adds the 2,3-DPG metabolic regulation of Hb affinity for O_2 to the physiological ΔpO_2 between lung and cerebrospinal fluid (CSF). Outside the red cell Hb becomes characterized in the sole presence of O_2, characterizing only the homotropic interconversion in the changes of quaternary structure which in not associated to complexes formation.

The Hb potential to acquire metal chelating sites allows the molecular quaternary structures T- and R- to participate in 2,3-DPG-Hb and divalent metal-Hb complexes stabilizing changes in the O_2 affinity. These ones induced on Hb by endogenous heterotropic ligands within the red cells, allowing the adaptive response to the O_2 needs of tissues and fulfilling in brain the matching of the rates of glucose and O_2 uptake.

The homotropic tendency of isolated Hb shows that during oxygenation a link between His (histidine) β146 and Asp (aspartate) β94 breaks up and the imidazole ring at above pH=6 takes up a proton (alkaline Bohr Effect), which may be able to trigger the changes from T to R as a sole function of O_2 through a ΔH=3kcal/mol [145, 146].

The erythrocyte concentrations: 65μg/g for Mg^{2+} and 11μg/g for Zn^{2+}, lead to physiological significant forms of the coordinative complexes: $(Mg^{2+})_2$Hb, $(Zn^{2+})_2$Hb and (Mg^{2+}or Zn^{2+}) Hb stabilizing complexes of higher affinity for O_2 than the R-form of Hb [99, 138].

In figure 12 were reported plots [95, 99] both directed to use the cooperativity value "n" as exponential of the pressure of oxygen: $(pO_2)^n$. This treatment allows obtaining straight curves. These predicts that the extrapolation to a maximal $(pO_2)^n$ renders an homotropic dependence on the transition between T- and R-forms of Hb. However, it could be predicted that the physiological ligand-Hb equilibriums within the red cell, would lead to an heterotropic mutual inclusion of O_2 and divalent metals Zn^{2+} or Mg^{2+} with a mutual exclusion of 2,3-DPG and H^+. The exergonic association per atom of Zn^{2+} to Hb: ΔH=9.7kcal/mol [141], the tetramer has two binding sites for either Zn^{2+} or Mg^{2+}. In the direction of mutual inclusion of divalent metal and oxygen the exergonic binding of Zn^{2+} of 19.4kcal plus pH dependent neutralization of Bohr protons, exceeds the endergonic dissociation of 2,3-DPG from Hb of ΔH=5.7kcal/mol. In the lungs this thermodynamic coupling integrates pO_2 with other contributions for a decrease in free energy. At lower pH than 7, the increasing stability order: $(Mg^{2+})_2$-oxyHb < (Zn^{2+},Mg^{2+})-oxyHb < $(Zn^{2+})_2$-oxyHb, indicates pH-differentiable complexes with O_2 dissociation all along the body pH-axis.

Dynamics of Ligands Modifying the Equilibrium of R to T State of the Quaternary Structure of Hemoglobin

Functional magnetic resonance imaging (fMRI) relies on the fact that cerebral blood flow and neuronal activation are coupled and the technique offers a higher spatial and temporal resolution without using radioactive material. The % of oxygen saturation of blood can be measure because the paramagnetic properties of Hb-O_2.

Iron's oxidation state in oxy-Hb is diamagnetic (no net unpaired electrons), yet the low-energy electron configurations in both oxygen and iron are paramagnetic (suggesting at least one unpaired electron in the complex).

Covalent bonds are not required to have perfect bond orders involving whole electron transfer. By resonance the actual electronic configuration of iron in Hb-O_2 is closer to being represented by Fe(III) and in the deoxy form by Fe(II).

The deoxygenation of oxyHb implies the gain of one electron for reducing the iron atom.

The uptake of 2,3-DPG for transition of oxyHb into deoxyHb involves an increase in pKa, from 6.6 to 8.2. Hence, imidazole group increases H^+ association, affecting the 2,3-DPG-Hb complex through the partial sharing of 2,3-DPG apported negative group.

The tight contact between capillary and neuron allows a partial negative charge displacement into the electrogenic membrane, allowing to attract Na^+ ions, initiating voltage-gated input (figure 28) into the action potential.

The multiplicity of complexes allows that deoxygenation could occur over an extended pH-range, than could be possible for any particular divalent metal complex of Hb. The pH of blood remains 7.4 but surroundings tissues allow increasingly lower pH from the trunk to the limbs. Hence, deoxygenation could follow a pattern of gradual release of O_2, first from the Mg^{2+}-Hb and thereafter from the Zn^{2+}-Hb species. However, the augment of glucose level in blood in untreated diabetes induces in the erythrocyte an increase of sugar phosphates. These ones by releasing the divalent metal from its ligand state with Hb decrease the Hb molecule affinity for O_2.

A premature release of O_2 allows O_2-depleted Hb to reach the body limbs and subsequence tissue anoxia results in the ulcerations characteristics of untreated diabetes. In the red cell, at the dimerization interphases $\beta_2\alpha_1$ and $\beta_1\alpha_2$ of Hb, pO_2 is mutually inclusive with the heterotropic ligands Mg^{2+} and Zn^{2+}, and could changes the ions coordinative state from "bi-" to "tetra-", and "hexa"-dentate. The more stable geometric configuration could be an octahedron with the β_2 R-groups: His (histidine) 143, Cys (cysteine) 93, Asp 94, His β146 and the R-group His α87, at the α_1-Heme. A symmetric structuring of the R-groups at the interphase $\beta_1\alpha_2$ will result in the formation of an oxyHb$(Me^{2+})_2$ complex, in which the quaternary form of Hb would show a lower tendency for O_2 dissociation.

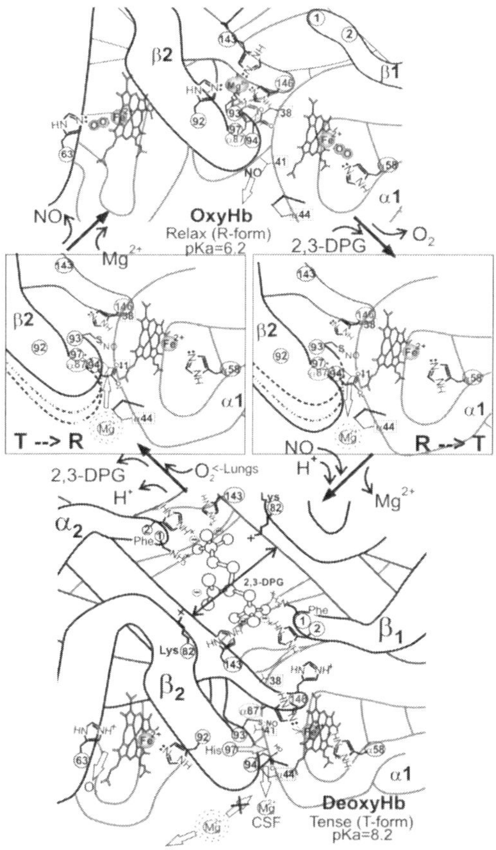

Figure 13. Conformational changes in the transition from oxyHb to deoxyHb. Inclusion of 2,3-DPG and H^+ couples with mutual deoxygenation and dissociation of Mg^{2+} or Zn^{2+}. The illustration has been scaled for Hb projected with a diameter of 5.5nm [139]. Freedom for Cα rotation of the side chains implicated for their participation on the coupling model described chelated structures, were tested with the generous assistance of Fermi, G. and Milstein, C. in the large tri-dimensional model of Hb, and located in the laboratory of M.J. Perutz, Cambridge, England. The figure shows only the $\beta_2\alpha_1$ interface, since structural symmetry allows an equivalent set of changes at the $\beta_1\alpha_2$ interface. Peptide bonds are rigid and fixed in a plane in where two Cα, 3.6Å apart, rotate 180o over angles ϕ (fi) and ψ (psi). The longitudinal lengths from Cα to their coordinating negative atoms: His=0.5nm, Cys=0.23nm and Asp=0.39nm; Magnesium Radius: atomic=0.150 nm and ionic (2+ ion)=0.086 nm; Zinc radius: atomic=0.135 nm and ionic (2+ ion)=0.088 nm; Iron radius: atomic=0.140 nm and ionic (2+ ion)=0.077 nm.

These events facilitate up to an 180° turn-around of the R-groups of Hb over their Cα. This effect adds to the directional position of histidines another 0.5nm length to reach the 1nm coordinative distance to the metal binding site. The chelating structure becomes conformational stabilized by the forming of the divalent metal-Hb complex. Simultaneously, at the lung pH=7.4 neutralization of H^+ released from the histidines of Hb induce a same charge rejection, favoring the release of 2,3-DPG from its state as ligand of Hb and increasing O_2-affinity allowing that the higher pO_2 by decreasing free energy favors the direction of oxygenation versus that of deoxygenation. At tissue level the change in free energy became associate to deoxygenation by the increase in the erythrocyte 2,3-DPG concentration and the much lower pO_2 [147]. Figure 13 illustrates a model of mutual inclusion of divalent metals and O_2 for the dimer $\beta_2\alpha_1$. The symmetric dimer $\beta_1\alpha_2$ do not have to be illustrated to explain mutual inclusion, but fully oxygenation involves saturation steps from mono to the tetra oxygenated molecule: $HbO_2 \rightarrow Hb(O_2)_2 \rightarrow Hb(O_2)_3 \rightarrow Hb(O_2)_4$. Each state of oxygenation, coupled by the concerted interaction between the two divalent metals ligand sites, in which the coordinative sharing of His $\beta_2$143 and His $\beta_1$143 are being subtracted step by step. This effect jointly with the displacement of the other ligands, which were conforming the binding site of single 2,3-DPG allow for the gradual disappearance of an structure capable to bind 2,3-DPG. Hence, the binding of each O_2, decreasing the free energy requirements for the binding of the subsequent oxygen.

Peptides bonds form a rigid plane in which the two Cα are free to rotate 180°. Coordinate bonding to Mg^{2+} allows a chelate between His β92 N:Mg^{2+}:N His $\beta_1$143+2H^+. This conformational change mutually increases affinity of Hb for Mg^{2+} and O_2 to form oxyHb, the two His β92 sequentially move away from their Heme to a position favoring O_2 binding to the respective Heme and β143 by moving away closes the 2,3-DPG binding site. At the cerebrospinal fluid (CSF) pH=7.4, oxyHb pKa=6.6 release O_2 and capture 2,3-DPG to transform the R (relax) into T (tense) form of deoxyHb pKa=8.2.

Mg^{2+}, O_2 and the hydration shell of 60 H_2O molecules are mutually inclusive ligands during Hb oxygenation, which losses 60 H_2O molecules into CSF during deoxygenation.

The turnover to deoxyHb became exergonic by coupling with H-bonds decrease in water clustering size. The latter, conveyance to the outside of the body implicates an open system with the role of releasing entropy. Homeostatic temperature characterizes the steady state for entropy exclusion.

The formation of the Mg^{2+}-oxyHb complex not only involves an Mg^{2+} transport mechanism, but also links the availability of O_2 to the activatory effects of Mg^{2+} on many enzymes. Moreover, the simultaneous increase in Hb affinity for both O_2 and Mg^{2+} involve coordinative bonds between Histidine residual R-groups of Hb with the tendency to chelate Mg^{2+}, but at the expense of the partial loss of the hydration shell of Mg^{2+} or Zn^{2+}. At homeostatic pH=7.4 glucose uptake by the erythrocyte increases the intracellular 2,3-DPG favoring transition from oxy to deoxyHb.

Cα-rotation dynamics allows that one or two Mg^{2+} at the $\beta_2\alpha_1$ and similarly at the $\beta_1\alpha_2$ symmetric crevice of these dimers of Hb could form tetradentate chelates. At the homeostatic neutral pH of CSF the imidazole rings His β92 and His β143 when coordinated by Mg^{2+} and/or Zn^{2+} in the Hb conformational sense deoxyHb pKa=8.2 to oxyHb pKa=6.4 [95, 99, 148].

After mutual inclusion of O_2 and Mg^{2+} stabilization of the conformational change into the R (relax) form $[(O_2)_4Hb(Mg)_2].(H_2O)_R$ reverse the sliding and in oxyHb two proline α 44 moves to close both gaps.

The tendency for mutual inclusion of O_2 and divalent metal allows a spectrum from greater to lower affinity of oxyHb chelated complexes $(Zn^{2+})_2(O_2)_4Hb$, $(Mg^{2+}/Zn^{2+})(O_2)_4Hb$, $(Mg^{2+})_2(O_2)_4Hb$ and $(O_2)_4Hb$. This allows that O_2 could not be released from Hb at a narrow pH peak, physiologically preventing a full discharge of O_2 by Hb, before reaching tissues at lower pH. Untreated diabetic patients show anoxia producing vascular necrosis, a diet which Zn^{2+} could increases the proportion of Zn^{2+} and greater peripheral tissue oxygenation.

The Zn^{2+}/Mg^{2+}-His chelation sites is a mechanism which has been under the denomination of Zn^{2+}-fingers, applied to many proteins and to a family of cellular nucleic acid binding proteins (CNBPs) [149, 150]. A divalent metal vs dynamics of protein changes in conformation can play the role of switches able to provide conformational mechanism for variable affinity turnover. It could be of biological importance by regulating the reversal of an enzyme fitting order with the tendency to favor substrates fitting decreasing after conformational changes increasing facility for products release. Thus, increasing an enzyme catalytic efficiency, or control of turnover through signaling like Mg^{2+} vs Ca^{2+} capable to change the binding constant between a receptor and its ligand hormone.

A Mechanism for Hb-Deoxygenation at Neutral pH

By incubating cells in media containing acetate, pyruvate, etc., the total concentrations of acid-soluble organic phosphate accumulation fractions depends on the extracellular phosphate level [101, 102, 103, 151]. Human erythrocyte incubated with pyruvate and inosine show that the total concentration of acid-soluble organic phosphates increases from 14 to 55 μmol/g red cells. Organic phosphate excluding 2,3-DPG can be elevated in vitro to 24 μmol/g.

The cerebrospinal fluid (CSF) pH=7.4 prevents pH-changes at the brain level, but the red cell heightens the concentration of the sugar phosphate intermediates of anaerobic glycolysis. The dynamics of divalent metals association with exclusion of 2,3-DPG in Hb and viceverse. The low pH inhibits 2,3-DPG mutase, allowing anaerobic glycolysis without generation of 2,3-DPG.

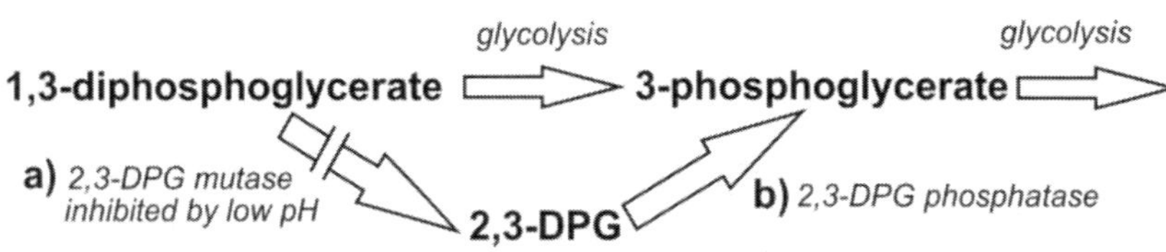

Scheme 1. The 2,3-DPG Rapapport-Luebering pathway. a) diphosphoglycerate mutase transfers from a carbon one (C1) phosphoryl group of 1,3-DPG to C2 of 2,3-DPG that by action of b) diphosphoglycerate phosphatase forms 3-PG. In the erythrocyte pH is an inhibitor of 2,3-diphosphoglycero mutase. The insulin-independent uptake of glucose, supports anaerobic glycolysis in the erythrocyte, allowing maximal activity of 2,3-DPG mutase, at the cerebral spinal fluid (CSF). The concentration of 2,3-DPG became maximal at pH=7.4 because 2,3-DPG mutase becomes inhibited by low pH allowing the shift the from oxy- to deoxy-Hb.

The insulin-independent glucose uptake of erythrocyte allows its endogenous sugar concentrations to be directly related to that in blood. Anaerobic glycolysis within the erythrocyte could generate 2,3-DPG to reach is highest endogenous concentration, because the pH inside the blood capillaries surrounding the CSF, would be close to the latter homeostatic-pH 7.4. The lack of a pH differential across the BBB indicates that the displacement of the equilibrium from the oxy to deoxyHb for the continuous release of O_2 into CSF, becomes only dependent in the increment in 2,3-DPG. The binding of 2,3-DPG of $\Delta H=-5.7$kcal/mol favoring the release of O_2 to match oxygenation for a sustainable aerobic glycolysis of the activated brain areas.

The control of the erythrocyte 2,3-DPG level fulfills, the role of a glucose level sensory capability of the Hb-red cell system. The molecular dynamics proposed allows that the sharing of some of the Hb R-groups allow binding of O_2 and the activatory ligands Mg^{2+} or Zn^{2+} with exclusion of 2,3-DPG or viceverse.

The energy necessities of the red cell, devoid of mitochondria, indicate the important role of anaerobic glycolysis. The regulation of oxygen-hemoglobin erythrocyte-system have requirement for both NADPH produced by the hexose monophosphate pathway (HMP) and 2,3-DPG the most concentrated organophosphate which varies inversely with pH [152].

Cyclic adenosine monophosphate (cAMP) [153] and cyclic guanosine monophosphate (cGMP) [154] are transported into human erythrocyte (Human-RBC) against significantly large differentials of the intracellular-extracellular concentrations [155, 156]. 0.1mM Dibutyryl-cAMP and dibutyryl-cGMP, compounds were found to diffuse freely into the erythrocyte. However, even if cAMP and cGMP catabolism did occur, were not able to elevate endogenous lactate and 2,3-DPG. Absence a guanylyl or adenylyl cyclase activity in the erythrocyte or the presence of cyclic nucleotides signaling was shown [157]. However, at 0.1μM dibutyryl-cGMP inhibits lactate production, which could be a mechanism contributing to pH homeostasis at the BBB.

Control of metabolic shifts on the erythrocyte appears to rest mainly in the interconnection of glycolytic pathways with the catabolism of cyclic nucleotides. The HMP-pathway production of CO_2 is decreased by cAMP and cGMP. Effect prevented by theophylline-dependent inhibition of phosphodiesterase, the enzyme producing AMP and GMP. Inhibition of glycolysis occurs as a feedback by the mass action of ribose or ribose phosphate, as a product buildup from the oxidative branch of the HMP pathway. This increases in the non-oxidative branch 2,3-DPG levels. Inosine, could also provide relatively small increases in the pathway of 2,3-DPG [154, 155, 158, 159].

Human-RBC, hemolysates show an increment of $^{14}CO_2$ released from the consumption of [1-^{14}C]-glucose within the HMP pathway. The Mn^{2+} stimulation of the redox recycling of NADPH contributes to pH-homeostasis because otherwise could favor lactic acid formation [160].

The Role of BBB in Modulation of Red Cell Control of Hb Deoxygenation

In most of the body, a network of vessels carry lymph, a fluid that removes excess plasma, dead blood cells, debris and other waste. But the brain is different. Instead of lymph, the brain is bathed in cerebrospinal fluid, which was assumed that by slowly diffusing through tissues, could be a waste carrier from the nervous system into the body's bloodstream.

The blood-brain barrier (BBB) is characterized as a dynamic frontier with cells at both sides exchanging nutrients across molecular gates and white blood cells crossing for protection. The cerebrospinal fluid (CSF) bathes the subarachnoid space and the ventricular system, around and inside the brain and spinal cord.

The endothelial cells (ECs) of the arteriole microvasculature of brain form tight junctions (TJs) that act as a diffusion barrier. Pericytes are mesenchymal cells that can modify their ultrastructure and gene expression in a unique synergistic relationship with the capillary ECs in the regulation of permeability.

The BBB shows capillaries without fenestrations or pinocytotic (transportation) vesicles, and have tight junctions that almost fuse adjacent cells.

Efficient mechanisms guard the constancy of the CSF pH 7.30-7.36 units, even in non-respiratory and to some extent in respiratory acid-base disturbances.

The pH of blood is maintained around 7.4, and buffering is required to maintain homeostasis. The blood contains Na^+, K^+, Ca^{2+} and Mg^{2+} cation, and Cl^-, PO_4^{3-} and SO_4^{2-} anions, which play a role in buffering. In surrounding peripheral tissues pCO_2 is about 50mm Hg, which allows diffusion into plasma and erythrocytes at about 40 mm Hg.

Into the erythrocyte, the H^+ dissociated from HCO_3^- ion is buffered by hemoglobin. The production of H^+ within erythrocytes and their subsequent buffering by hemoglobin, results in a reduced affinity of hemoglobin for oxygen. This leads to a release of O_2 to the peripheral tissues, a phenomenon is termed the Bohr Effect, which is potentiated by the change in the endogenous 2,3-DPG concentration in the erythrocyte.

The presence of chlorides ions: 120-130 mEq/l may not only play an isotonic role, because across the CSF barrier may help dissociation of the divalent metals from oxyHb and facilitate deoxygenation. Cerebral blood flow may be constricted at high CSF pressure. Disorders of CSF flow may also affect the intracranial blood flow with subsequent neuronal and glial vulnerabilities.

CSF contains < 10-12 U/ml of adenosine deaminase that breaks down adenosine into inosine and ammonia. Adenosine has inhibitory effects over neuronal activity, sedates and increments the non-rapid eye movement sleep (NREM) in stage IV. Caffeine blocking the adenosine binding to the receptor decreases sleeping tendency.

Lipophilic substances heroin, alcohol, cocaine, etc., and molecules smaller than 500 M.W. like sleep aids, antipsychotics and antidepressants, usually could cross the barrier as well as cortisol.

L-dopa ((S)-2-Amino-3-(3,4-dihydroxyphenyl)propanoic acid) could cross the BBB allowing use for Parkinson treatment, but adrenaline ((R)-4-(1-Hydroxy-2-(methylamino)ethyl)benzene-1,2-diol) could not cross. Thus, restricts its use to metabolic and bronchodilation effects on organs devoid of direct sympathetic innervation.

The proteins known as RAGE conduct the β-amyloid into the brain from the bloodstream and the protein LRPI blocks, if the balance becomes broken Alzheimer's plaques increase.

Chapter 5

Enzyme-Membrane Systems

From rat's tissues fat cells [6, 8] and brain (hypothalamus, cortex, and corpus striatum [1, 2, 3,4]) were obtained purified enzyme-membrane particulate preparations of AC.

The preparations of AC [ATP pyrophosphate-lyase (cyclizing) (EC 4.6.1.1)] obtained from corpus striatum, hypothalamus and cortex of the brains of rats show that noradrenaline activation is dependent of the concentration of magnesium ion (Mg^{2+}).

The insulin receptor tyrosine kinase (IRTK) was obtained from rat liver [14]. The membrane purification treatments preserved the allotopically association between enzymes and hormonal receptors. Physiological interaction is could be preserved by not extending purification to the separation of the enzyme from receptor and/or the membrane [10]. Enzyme-membrane particulate systems were shown to preserve the modulatory responsiveness to hormonal or neurotransmitters signaling.

Mg^{2+} Treated-AC Particles Preparation from Cerebral Cortex

Mg^{2+} treated particles were prepared as described under Methods to assay the Mg^{2+}-dependent stability of the adrenergic receptor coupling the enzyme. Portions containing the amount yielded by 20mg of rat cerebral cortex were incubated at 15°C for 10min in a final 1ml reaction mixture containing 40μmol of Tricine-Tris buffer, pH=7.4, 6.67μmol of caffeine, ATP-Mg adjusted with

Tris to pH 7.4 at the concentrations indicated in the figure, in the absence or presence of 1µmol of noradrenaline (NA). The reaction was terminated by heating at 100°C for 5min.

Reaction blanks were done by mixing a volume of the particle suspension with the other components of the reaction mixture and heating immediately in a boiling-water bath. The mixtures were centrifuged in a GSA head of a Sorvall RC2 (10°C) at 4500rev/min for 15min. The supernatants were collected and 50µl portions were assayed for cyclic adenosine monophosphate (cAMP) by the procedure of Brown et al. (1972) [9].

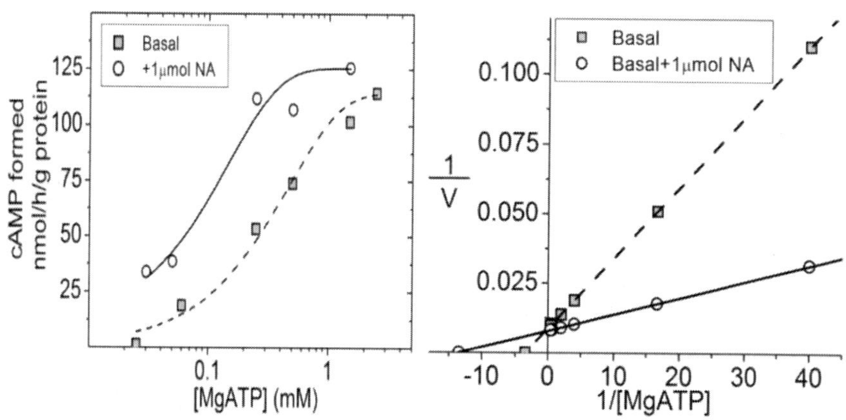

Figure 14. a) b) c) Effect of noradrenaline (NA) on a substrate saturation curve of Mg^{2+}-treated particles from cerebral cortex. After preincubation, the final pellets were suspended in 1.2ml of the same buffer, and 0.25ml portions were incubated in a final 1ml volume containing 0.25mM-ATP (Sigma: sodium-ATP from horse muscle containing 2% of GTP and minimum quantities of other nucleotides) in the absence or presence of 1mM-NA in the same reaction mixture and conditions. Specific activity is expressed as nmol of cyclic adenosine monophosphate (cAMP) formed/h/g of cerebral cortex used to prepare the particles.

cAMP formed was calculated from a calibration curve. The results were expressed as specific activity of adenylate cyclase converting values for wet wt. by multiplying 4.3 into equivalent mg protein (nmol of cAMP formed/h per of cerebral cortex used to prepare the particles).

NA appears to play a rather specific role in sensitizing a labile enzyme to environmental temperature-dependent effects that affect the enzyme-membrane structure required for AC activity. Since substrates have been found to protect some enzymes, MgATP may decrease the effect of NA by stabilizing the structure of AC, thus decreasing the sensitivity of the NA ligand-enzyme membrane complex as a structure subject to heat inactivation at room temperature.

Table 1. Effect of preincubation on the stability of Mg^{2+}-treated particulate adenylate cyclase of cerebral cortex.
Portions of 120mg of rat cerebral cortex were preincubated at 38°C for 3h and 20min in a final 3ml suspension of 2.8mM-Tricine-Tris buffer (pH=7.4)-0.25mM-$MgSO_4$ and the final concentration of the addition indicated in the table. It shows that only after 3h of preincubation in buffer at 38°C did the Ca^{2+}-treated particles lose 50% of their adenylyl cyclase activity

	Specific activity of adenylyl cyclase nmol/h per g		Adenylyl cyclase activity as % of control	
Preincubation additions	-NA	+1mM-NA	NA-independent	NA-dependent
5mM-ATP-Mg	30.3	25.8	252	-14.8
0.1mM-NA	0.0	0.0	0	0
5mM-ATP-Mg+0.1mM-NA	10.8	6.3	90	-42
Buffer alone	12.0	35.6	100	+196

Table 2. Effects of Mg^{2+} and ATP^{4-} on the response to noradrenaline and calcium of adenylyl cyclase of corpus striatum. To investigate the individual and separate effects of free Mg^{2+} and of ATP^{4-}, their concentration were determined by multiple-equilibrium equations

Kinetic Constants	Basal	0.1mM-NA	0.1mM-NA+0.3mM $CaCl_2$
Mg_T	n=2.7	n=4.4	n=2.3
Mg_T	$S_{0.5}$=8mM	$S_{0.5}$=9mM	$S_{0.5}$=9.4mM
Mg^{2+}	n= 1.1	n=1.9	n=1.3
Mg^{2+}	$S_{0.5}$=4.7mM	$S_{0.5}$=4.6mM	$S_{0.5}$=4.6mM
ATP^{4-}	n=1.1	n=1.9	n=1.3
ATP^{4-}	$S_{0.5}$=0.1mM	$S_{0.5}$=0.1mM	$S_{0.5}$=0.1mM

The AC of corpus striatum, at basal conditions and for 0.1mM-NA and +0.1mM-NA+0.3mM $CaCl_2$, shows similar affinity $S_{0.5}(Mg_T)$ and $S_{0.5}(Mg^{2+})$ of about 9mM. The affinity for the inhibitory specie $S_{0.5}$ (ATP^{4-})=0.1mM.

NA mainly increases Vmax but does not change by much the enzyme affinity for either Mg^{2+} or ATP^{4-}.

The AC activity of corpus striatum for 0.1mM-NA cooperativity n=4.4. Cooperativity for basal n=2.7, since for basal n(Mg^{2+}), n(ATP^{4-}) and n(MgATP) show values close to 1.

The values obtained for basal cooperativity n(Mg^{2+}), are influenced by the summation to the activatory effect of Mg^{2+}, and the disappearance of the inhibitory effect of ATP^{4-}.

The enzyme saturation curve shows a n(adrenaline) values of about 4, because the neurotransmitter potentiates Mg^{2+} mediated binding to the membrane. The hormonal receptor for the expression of properties dependent of its position in the membrane, allotopic, may require Mg^{2+} [10] for the activatory binding of G-protein. The enzyme shows a single active site, therefore, requires Mg^{2+} and MgATP for basal activity, and a separate for the binding of an adrenaline receptor.

The applicability of the random of the RARE BiBi model is confirmed by the ability of the model to predict a change of Hill cooperativity from a basal value of n=2.6 to 4.0 in the presence of NA. This indicates that excess ions Mg^{2+} over that required for the saturation of sites for substrate formation and for the Mg^{2+} regulatory activation. An additional saturation by ionic Mg^{2+} is an obligatory requirement for the manifestation of third site linked to the responsiveness to hormonal dependent activation [2, 6, 8].

EDTA Treated-AC Particle Preparation from Cortex, Striatum and Hypothalamus

EDTA treated particle were prepared as indicated in the Methods to differentiate the effect of Mg^{2+} vs Ca^{2+} on the responsiveness to noradrenaline (NA) of the α_2 adrenergic receptors. The activities of EDTA-treated particle preparation of adenylyl cyclase from cortex, striatum and hypothalamus were assayed.

Additions as indicated in the figures and table 3. Values for enzyme activity: assays-(0-time blanks) reported in the vertical-axis. The incubation was terminated by heating. Each curve was plotted using dots averaging 4 assays. Vmax values are shown with the standard error of the mean (SEM).

Table 3. Kinetic characterization of adenylyl cyclase obtained from specifics brain areas. a) EDTA-particles containing 100μg of protein determined by the colorimetric procedure of Lowry et al., (1951) were incubated at 25°C for 60min in 40mM-Tricine/Tris buffer, pH=7.4, 6.67mM-caffeine, constant 1mM-ATP

Particulate Membrane preparations	Km(MgATP) [mM]	Cooperativity (n)
Cortex	2.2	1.4
Striatum	2.0	1.1
Hypothalamus	2.0	1.1

The values obtained do not allow inferring differences between the responsiveness to the substrate of the basal adenylyl cyclase activity of the three brain areas studied.

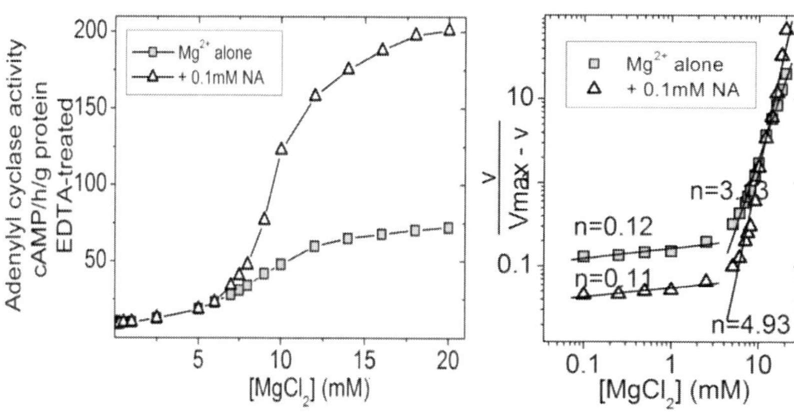

Figure 15. a) Effect of noradrenaline (NA) on the activity of EDTA-treated AC particles from cortex. The stimulatory α-subunit (Gαs) of the heterotrimeric-G-protein with Mg^{2+} activates adenylyl cyclase. ◻) Mg^{2+} alone, △) +0.1mM NA, ○) +0.1mM NA +0.3mM $CaCl_2$.

An addition at the indicated concentrations: in a) ◻) $MgCl_2$ alone (control), Vmax 76±7, and △) control + 0.1mM NA, Vmax 204±14; in b) ◻) $MgCl_2$ alone and ○) control + 0.1mM NA + 0.3mM $CaCl_2$, Vmax 27±2.

The observed effect of a Mg^{2+} activatory effect vs a Ca^{2+} inhibitory effect beyond that of the basal enzyme responsiveness was attributed to the iso-

subunit Gαi binding AC in the presence of Ca^{2+} to activate an inhibitory α-receptor.

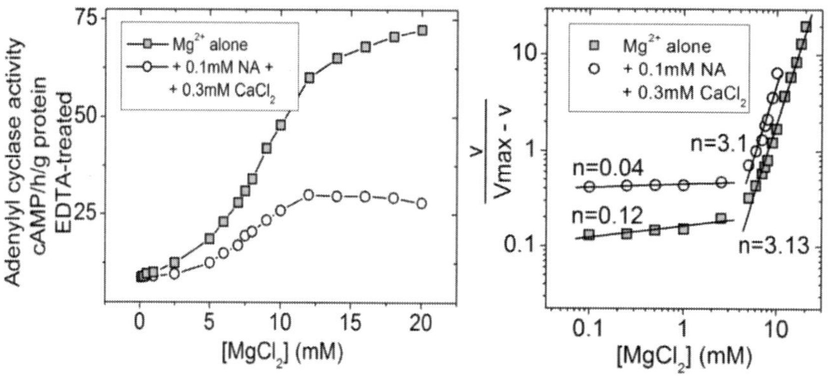

Figure 15. b) Effect of NA on the activity of EDTA-treated AC particles from cortex. The inhibitory α-subunit (Gαi) of the heterotrimeric-G-protein with Ca^{2+} shows the characteristics of an inhibitory receptor coupling with AC.

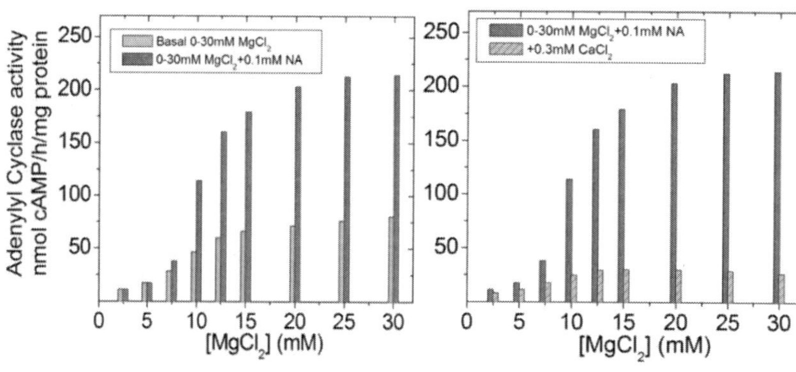

Figure 16. The effect of Mg^{2+} vs Ca^{2+} on the modulation of neurotransmitter responsiveness of AC. The EDTA-treated-particle preparations from rat brain cortex in samples containing 100μg of protein by the method of Lowry et al., were assayed for AC-activity by incubation at 25°C for 60min in 40mM-Tricine/Tris buffer, pH=7.4, 6.67mM-caffeine, constant 1mM-NaATP (Sigma: horse extracted containing 2% of GTP and other nucleotides).

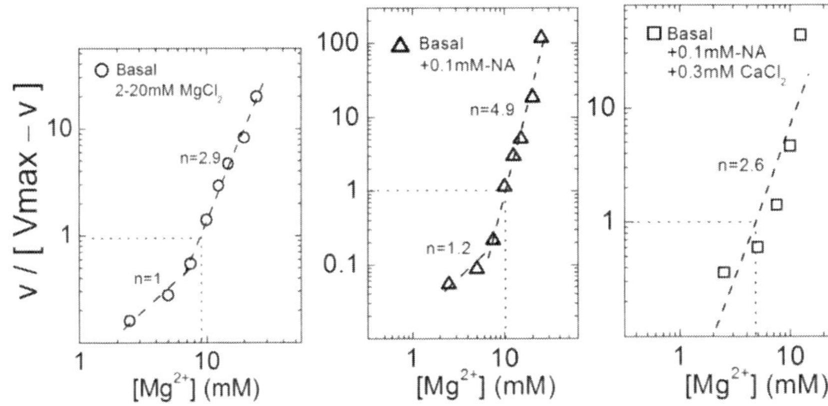

Figure 17. Hill plots of $MgCl_2$ saturation curve of AC preparation from cortex at constant 1mM-NaATP. O) basal 0-30mM $MgCl_2$, Δ) 0-30mM $MgCl_2$+0.1mM NA, □)+0.3mM $CaCl_2$.

Table 4. Noradrenaline responsiveness of AC-cortex. Activity: nmol of cyclic adenosine monophosphate (cAMP) formed /h per mg of purified protein as Ethylenediaminetetraacetic acid (EDTA)-particulate extract from rat cerebral cortex. These were incubated with varying concentrations of Mg^{2+} and 1 mM MgATP and 0.02 mM GTP (Sigma purified equine muscle), value of cooperativity Hill, n = 3.9 [2]

Divelent metal	Basal	+0.1mM NA	Δ% NA vs basal
15mM Mg^{2+}	78±7	212±14(4)	+172
25mM [$MgCl_2$]+0.3mM Ca^{2+}	0	27±2(4)	0

Mg^{2+} Requirement for Noradrenaline

Adrenergic nerve fibers make up the sympathetic nervous system which releases the neurotransmitter norepinephrine (NE) at the synapse, or junction. Alterations in the endogenous cyclic AMP could be caused by changes in metabolic conditions for example, as the inhibition of phosphodiesterase-dependent breakdown of cyclic AMP. The isolation from rat's brain of adenylyl cyclase (AC) [EC 4.6.1.1] and characterization for its neurotransmitter responsiveness [2] allows evaluation of a possible role of the enzyme in cyclic adenosine monophosphate (cAMP)-dependent memory processes. NE activation, when Mg^{2+} is in excess of substrate of an isolated AC from brain, was first reported for cortex and corpus striatum [4]. The enzyme relates to a network of tissues [6, 161] supporting brain activity [162] by their *in common* response to the modification of free ionic Mg^{2+} by changes in the concentration of chelating metabolites.

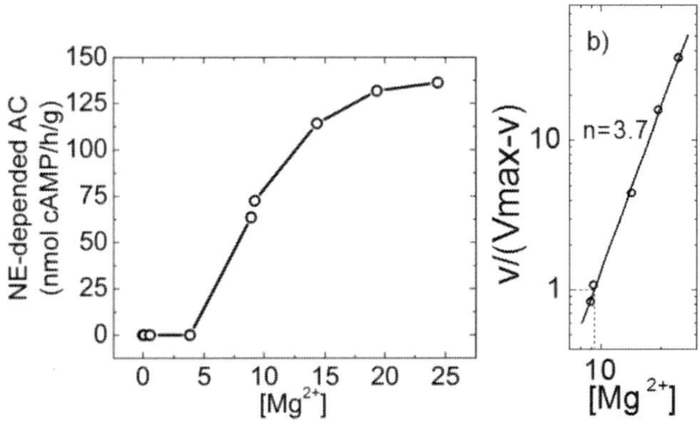

Figure 18. Differential of noradrenaline (NA)-dependent vs Mg^{2+}-dependent stimulation of AC preparation from cortex. 100μg of protein from the EDTA-AC-membrane preparation in 40mM-tricina/Tris-buffer, pH= 7.4, 6.67mM-caffeine and constant 1mM ATP (Sigma: purified horse muscle containing 2% GTP) and addition of $MgCl_2$ for a saturation curve in 0.5ml final volume were incubated at 25°C for 1h in the presence of 0.1mM-norepinephrine (NE Vmax=214) and its absence (basal Vmax=76±7(4)). Plotted data o) the difference between: basal+0.1mM-norepinephrine and basal curves, at constant ATP^{4-} concentration.

The purpose was to evidence the integration of neurotransmitter responsiveness of AC with the regulatory feedbacks of the metabolic network, supporting neuronal aerobic glycolysis. $MgCl_2$ saturation curves of AC show cooperative interactions at three sites.

An obligatory free Mg^{2+} requirement for the enzyme responsiveness to noradrenaline (NA) with a cooperative value n=4 and 300% increase of Vmax. The NA effect may mediate an emotionally activated long-term memory, which requires a sudden increase of glucose and oxygen consumption.

Previous kinetic studies of the NA-activated adenylyl cyclase (AC), evidenced that the neurotransmitter response could be differentiated from a basal response, according to their different Vmax, which could explain memory at two differentiable levels. The NA-dependent activity of AC could be induced by events with strong emotional connections, related to the fight-or-flight response.

Kinetic Properties Relevant to Emotional Reinforcement of Basal-AC Memory Processes

Memory affirmation may require a restless basal AC-activity: Km(MgATP)=1.05 mM. The active site requires an excess of free Mg^{2+} over substrate to prevent dead end inhibition by ATP^{4-} and/or CaATP.

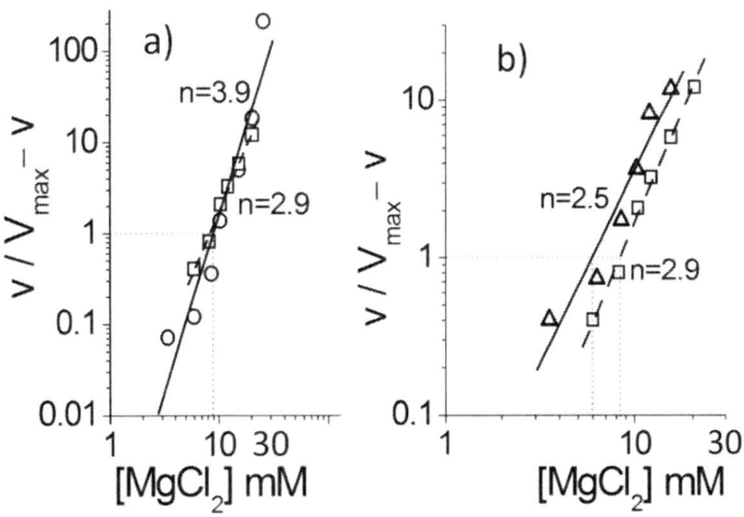

Figure 19. Control by MgT of the responsiveness of noradrenaline (NA) activated AC. Samples of an EDTA-particles preparation from cortex containing 100□g of protein were incubated at 37°C for hour, in 40mM-Tricine/Tris buffer, pH=7.4, 6.67mM-caffeine, constant 1mM-ATP(Sigma purified from horse muscle), in a final 0.5ml with the indicated concentration of $MgCl_2$ (MgT). Vmax for each curve was determined at the indicated $MgCl_2$ concentration. Activity reported: nmol of cyclic adenosine monophosphate (cAMP) formed/h per mg of membrane protein. Figure a). $MgCl_2$ saturation curves: o) +1mM-NA, at 30mM [MgCl2] were determined Vmax values, 212 ±14(4);Δ) basal, 78±7(4). Figure b) MgCl2 saturation curves: basal, at 15 mM [MgCl2], 78±7(4); □) +1mM-NA +0.3 mM CaCl2 at 25mM [$MgCl_2$], 27±2(4).

The addition of the basal Hill constants for Mg^{2+} and for ATP^{4-} adds to an n value of 2.4, which closely correspond to separated values.

This mechanism fits the finding of an additive cooperativity in the response of the enzyme, toward its simultaneous dependence of an increase in Mg^{2+} and decrease of ATP^{4-} concentrations.

Accordingly, the $n(Mg_T)$ for Mg^{2+} cooperativity in the presence of the tested hormones reveals two additional sites, which exceed the two regulatory sites, characterized for basal. This indicates that the Mg^{2+}-magnification stimulus on AC, relates to the interaction between catalytic and regulatory sites within the enzyme molecule itself, but also shows the presence of additional Mg^{2+} binding sites. Thus, indicating the presence of Mg^{2+}-dependent saturation requirements for the binding of AC and hormonal receptors into the isolated membranes.

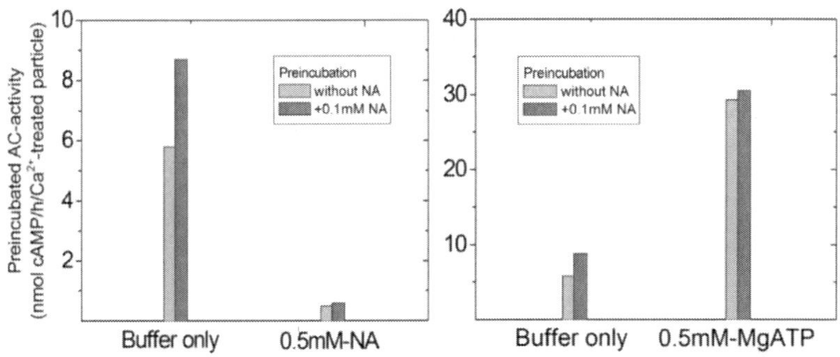

Figure 20. Destabilizing effect on adenylyl cyclase of brain cortex by overexposure to noradrenaline (NA). Samples from 36mg fresh tissue weight of Ca^{2+}-treated particle from isolated membranes of rat brain cortex in 10mM Tricine/Tris at pH 7.4, in a final volume of 0.9 ml, and the additions indicated in the figure, were preincubated at 38°C during 3 h as an assay of enzyme stability. The preincubation mixtures were centrifuged at 1700rpm from 15 min in an IEC model 809 head at 20°C. The pellets were washed twice, suspended in 3ml of the same buffer. Samples of preincubated particles, in 0.36 ml of the same buffer corresponding to 15mg fresh weight of tissue were incubated in a final volume of 0.5 ml. The residual adenylyl cyclase activity was measured in 1mM ATP free of GTP in the absence of NA or basal conditions light bars and with the addition of 0.1mM noradrenaline dark bars [2].

The results of the experimentally obtained prolonged association of noradrenaline (NA) with its receptor may correspond to a stressing of the quaternary structure, during the activatory state [96]. This catecholamine-dependent inactivatory event was prevented by a pre-incubation condition, in which the presence of MgATP protects from NA, by the formation of an enzyme substrate complex. The NA-AC complex manifests thermic instability, whereas the NA-AC-MgATP allows to protect the enzyme activity, but to the cost of decreasing its responsiveness to a catecholamine activatory effect [2, 96, 163].

The metabolic effects of ionic equilibrium are similar to an amplification system, since large increases in the reaction rate of AC, occur by a change in chelating metabolite concentration like ATP^{4-}, $citrate^{3-}$, etc. This could lead to the capturing of free Mg^{2+} or releasing it, and also introduce large changes in the concentration of the inhibitory species ATP^{4-}, CaATP, Ca^{2+}, etc. The role of adrenaline in mobilizing fat stores could be replaced by other hormones over a different range of $[Mg^{2+}]$. If ATP^{4-} became exhausted increases the equilibrium level of free Mg^{2+}, allowing adrenocorticotropic hormone (ACTH) to become activatory at a much lower $[Mg^{2+}]$ concentration ranges, than the one for physiological activation by adrenaline and glucagon.

Table 5. Kinetic constants of Mg^{2+} and ATP^{4-} on the responsiveness to noradrenaline (NA) and calcium by AC from corpus striatum.

The separate effects on AC by the individualized ions concentration were determined by multiple-equilibrium equations

Kinetic Constants	Basal	0.1mM-NA	0.1mM-NA +0.3mM $CaCl_2$
Mg_T	n=2.7	n=4.4	n=2.3
Mg_T	$S_{0.5}$=8mM	$S_{0.5}$=9mM	$S_{0.5}$=9.4mM
Mg^{2+}	n= 1.1	n=1.9	n=1.3
Mg^{2+}	$S_{0.5}$=4.7mM	$S_{0.5}$=4.6mM	$S_{0.5}$=4.6mM
ATP^{4-}	n=1.1	n=1.9	n=1.3
ATP^{4-}	$S_{0.5}$=0.1mM	$S_{0.5}$=0.1mM	$S_{0.5}$=0.1mM

Table shows that allotopic enzyme-membrane properties are similar for the brain regions tested.

Figure 20 shows the activatory effect of NA versus basal on the AC of cortex increasing by 300% basal Vmax. The cortex enzyme shows cooperativity for total-magnesium in the presence of 0.1mM-NA with a value $n(Mg_T)=3.9$.

AC basal cooperativity decreases from $n(Mg_T)=2.9$ and by addition of NA plus $CaCl_2$ decreasing to $n(Mg_T)=2.5$ and decreased Vmax to 1/3.

Comparative results on the effect of NA in AC of corpus striatum are reported in table. The AC of corpus striatum at basal conditions and for 0.1mM-NA and +0.1mM-NA+0.3mM $CaCl_2$, show similar affinity $S_{0.5}(Mg_T)$ and $S_{0.5}(Mg^{2+})$ of about 9mM.

The affinity for the inhibitory specie $S_{0.5}(ATP^{4-})$=0,1mM. NA mainly increases Vmax but does not change by much the enzyme affinity for either Mg^{2+} or ATP^{4-}. The AC of corpus striatum cooperativity for 0.1mM-NA n=4.4. Cooperativity for basal n=2.7, since for basal n(Mg^{2+}), n(ATP^{4-}) and n(MgATP) show values close to 1.

The enzyme saturation system, shows a n(adrenaline) values of about 4, because the neurotransmitter potentiates Mg^{2+} mediated binding to the membrane.

Basal Activity of AC from Corpus Striatum

To study the basal activity of AC required to use chemically synthetized sodium-ATP sigma without contaminants or GTP. The values obtained for basal activity n(Mg^{2+}), are influenced by the summation to the activatory effect of Mg^{2+}, and the disappearance of the inhibitory effect of ATP^{4-}.

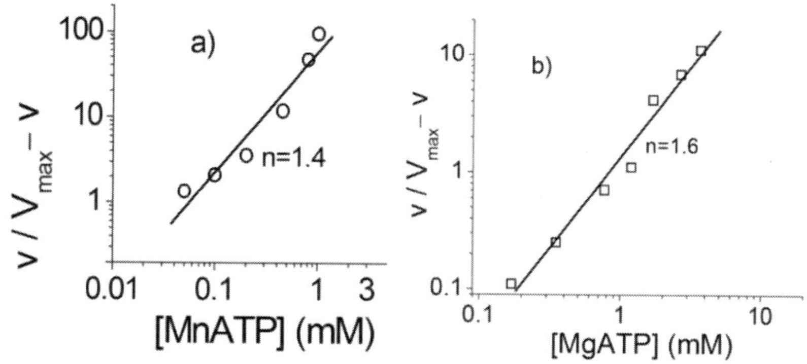

Figure 21. a) MnATP substrate saturation, and b) MgATP substrate saturation curves. Portions of AC particulate preparation from corpus striatum containing 6μg proteins were incubated in the conditions for assay of basal activity. Activity reported: nmol cyclic adenosine monophosphate (cAMP) formed/h per mg of membrane protein. a) MnATP saturation curve: o) At 1mM MnATP reached Vmax value: 19.2nmol cAMP, cooperativity value n(MnATP)=1.4, Km(MnATP)=0.05mM. b), MgATP saturation curve: At 3.6mM MgATP reached Vmax: 8.2nmol cAMP, cooperativity value n(MgATP)=1.6, Km(MgATP)=0.84mM.

The enzyme shows a single active site, therefore, requires Mg^{2+} and MgATP for basal activity, and a separate for the binding of an adrenaline receptor. Lineweaver-Burk plots of AC activity as a function of substrate concentration at different concentrations of Mg_{TOTAL} show a noncompetitive behavior, indicating that Mg^{2+} interacts with a distinct regulatory site differentiable from the active site [4].

Table 6. Equilibrium concentrations of MgATP, ATP^{4-} and Mg^{2+} under experimental conditions where ATP^{4-} is kept constant and MgATP is increased. Equilibrium concentrations, adjusted from the calculated ionic equilibrium [164, 165]

$[Mg^{2+}]$ (mM)	$[ATP^{4-}]$ (mM)	[MgATP] (mM)	$[MgATP]/[Mg^{2+}]$
0.0058	2.20	0.244	42.1
0.0229	2.23	0.976	42.6
0.0455	2.25	1.950	43.4
0.0896	2.29	3.910	43.4
0.1110	2.31	4.880	43.9
0.1320	2.33	5.860	44.0
0.2000	2.33	8.900	44.0

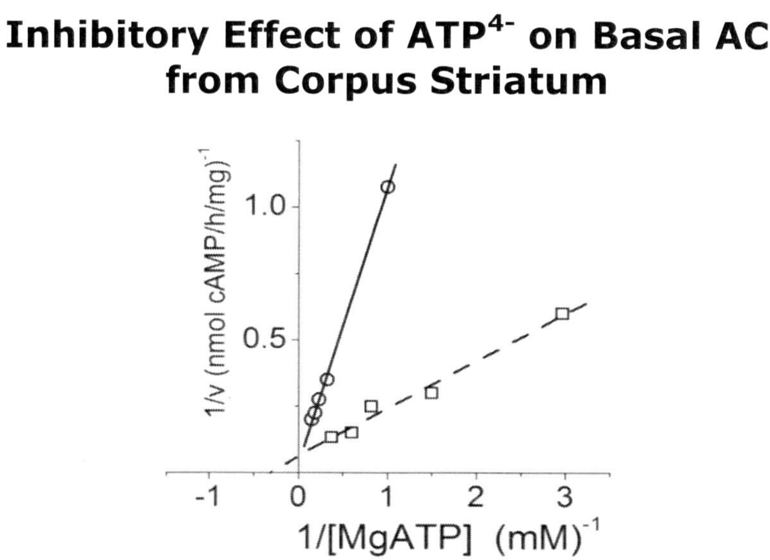

Figure 22. Lineweaver-Burke plot. The inhibitory effect of ATP^{4-} on the substrate saturation curve of AC particulate preparation from corpus striatum. Portions of an AC particulate preparation containing 2.5 μg of protein were incubated. □) Basal, MgATP saturation curve $K_{0.5}$=5mM MgATP, Vmax: 9nmol cyclic adenosine monophosphate (cAMP)/h/per mg of protein; o) basal+2.5 mM ATP^{4-} at $K_{0.5}$=9mM MgATP, $Vmax_{app}$: 7nmol cAMP/h/per mg of protein. Equilibrium concentration adjusted according to the above table.

The hormonal receptor for the expression of properties dependent of its position in the membrane, allotopic, may require Mg^{2+} for the activatory binding of G-protein.

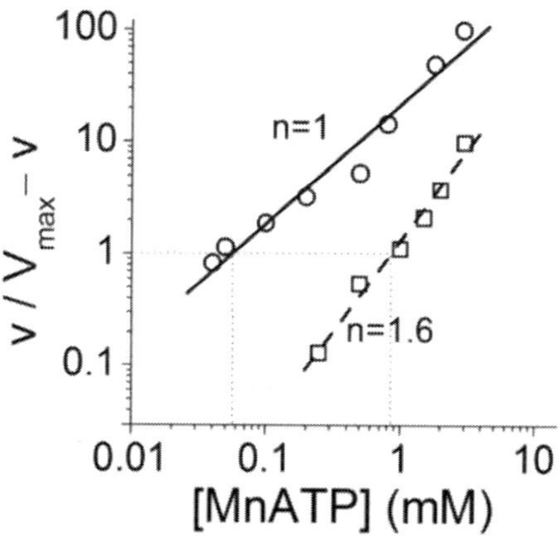

Figure 23. Effect of inhibitor ATP^{4-} on the substrate MnATP saturation curve of corpus striatum adenylyl cyclase. 6μg protein portions of an adenylyl cyclase-membrane preparation were incubated at 37°C. o) Basal; with the indicated increasing equilibrium concentration of MnATP at 1mM MnATP reached Vmax: 19.2nmol cyclic adenosine monophosphate (cAMP)/h/per mg; basal+inhibitor 5.5mM ATP^{4-} at 4mM reached Vmax: 15.2nmol cAMP/h/per mg.

The Hill Plot was obtained from saturation curves with a Vmax(MnATP): 19.2nmol cAMP/h/mg, a considerable higher activity than the corresponding Vmax(MgATP): 8.2nmol cAMP/h/mg.

Basal AC-activity shows that the substrate MnATP increases affinity Km(MnATP)=0.06 mM and 200% the Vmax. AC-dependent memory may have Mn^{2+}-activated switches for differentiable gating of nerve impulses, along specific synapses networks. Basal activity shows a cooperativity value, n=2.6, surging from the interaction between active and regulatory sites. The latter, affinity values: $K_{0.5}(E\text{-}Mg^{2+})=4.7$ mM, $K_{0.5}(E\text{-}Mn^{2+})=3.8$ mM and $K0.5(Ca^{2+})=0.02$ mM.

The increasing concentration of $[MnATP]/[Mn^{2+}][ATP^{4-}]=4\times10^4$, implies that equilibrium maintains negligible concentration of free Mn^{2+}. Hence, the addition to the basal curve of ATP^{4-} maintains its equilibrium concentration well above that in the basal curve. The ATP^{4-} inhibitory-curve approaches the basal at the higher maximal concentration, showing unchanged Vmax.

Kmapp(MnATP) changes from 0.075 to 1.5 mM . The calculated $Ki(ATP^{4-})=0.25$mM similar to the value obtained for substrate MgATP,

$Ki(ATP^{4-})$=0.27mM. Hence, the presence of MnATP does not alter the affinity for adenylyl cyclase for ATP^{4-}, ruling out, indirect modulation by Mn^{2+} on the affinity of the enzyme for ATP^{4-}.

Evaluation of the responsiveness of the enzyme indicates that there is a requirement that the concentration of free Mg^{2+} exceeds that of MgATP to activate the enzyme. The AC response appear coupled to the operation of the Na^{+} gates augmenting free ionic Mg^{2+}. The gates associated ATPase reaction releases free Mg^{2+} because since ADP^{3-} has a much weaker association constant for the divalent metal than ATP^{4-}.

Comparing both Km, it could be shown a marked increase for the basal activity of adenylyl cyclase in the presence of MnATP with an affinity increment of over 17 times than the Km MgATP. Moreover, Vmax was increased for the substrate MnATP 2.3 times over that of Vmax MgATP.

Equilibrium conditions for MnATP (log Ka=4.6) allowed to increase the concentration of MnATP at negligible concentration of free ionic Mn^{2+}. Hence, it could be inferred that if both substrates are separately compartmentalized the adenylyl cyclase by using MnATP would be able to produce cAMP at much higher rates than in the presence of MgATP alone.

The adenylyl cyclase in the presence of substrate MnATP at basal level generates cAMP independently of the already discussed free Mg^{2+} activatory requirement for responsiveness to the neurotransmitter. It could be inferred the existence of neurons or synaptic structures able to orchestrate in the absence of neurotransmitter a facilitated response of adenylyl cyclase, utilizing MnATP, which becomes supporting of AC memory networks required to have a limited sensitivity to changes in neurotransmitter levels.

Inhibitor Effect of CaATP on Basal-AC from Hypothalamus

The addition of CaATP produce competitive inhibition decreasing affinity from 2mM to 20mM K_{mapp}(MgATP).

Lineweaver-Burk, the addition of CaATP shows Ki=0.015mM. However, Mg^{2+} (Mg_T) can easily displace Ca^{2+} from ATP. The equilibrium favors that Mg^{2+} can easily displace Ca^{2+} from ATP, and reduce the concentration of CaATP. Since Ka (MgATP)=$-20\times10^{3}M^{-1}$ versus Ka(CaATP)=$-9\times10^{3}M^{-1}$.

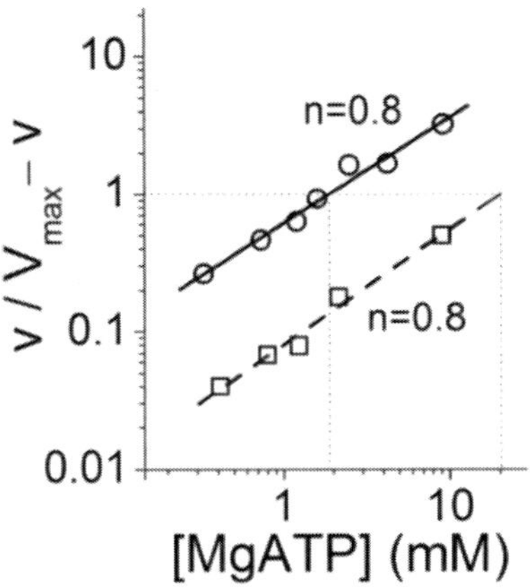

Figure 24. CaATP inhibitory effect on the substrate MgATP saturation curve of AC from hypothalamus. 3.3μg protein portions of a hypothalamic-membrane preparation were incubated at 37°C, increasing equilibrium concentrations of 0.41 to 9.3 mM MgATP: MgATP saturation curve: o) Basal at 9.3mM MgATP Vmax: 13.2nmol cyclic adenosine monophosphate (cAMP)/h/mg; □) basal+inhibitor 0.37±0.05mM CaATP at 9.3mM MgATP Vmax: 5.8nmol cAMP/h/mg. Cooperativity value: n.

The applicability of the random of the RARE BiBi model is confirmed by the ability of the model to predict a change of Hill cooperativity from a basal value of n=2.6 to 4.0 in the presence of noradrenaline. This indicates that excess ions Mg^{2+} over that required for the saturation of sites for substrate formation and for the Mg^{2+} regulatory activation. Hence, an additional saturation by ionic Mg^{2+} is an obligatory requirement for the manifestation of third site linked to the responsiveness to hormonal dependent activation [2, 6, 8].

The AC kinetics could be integrated in a random rapid equilibrium (RARE BiBi) with dead end inhibition Ki(ATP^{4-})=0.27 mM and Ki(CaATP)=0.01 mM. An increase in chelating metabolites increases both dead end inhibitors. Alternatively, a decrease in chelating metabolites decreases CaATP, strongly activating AC-mediated and cAMP-dependent activation of pathways for memory affirmation.

Table 7. Kinetic values from Hill and Lineweaver and thermodynamics constant for the indicated E-Complexes.

Kd dissociation constant; Kf formation constant for enzyme complexes; ΔG^o, Gibbs free energy. Kd values change by only 9% when ionic strength changes 100%. Hence, Kd values where affected by considerably lower deviations

Substrate Saturation	Additions	Enzyme Complex	Km (mM)	Kd (mM)	Ki (mM)	$Kf \times 10^4$ (M^{-1})	Cooperativity (n)	ΔG^o (kcal)
MnATP	---	E-MnATP	0.065	---	---	1.43	0.8	5.6
MgATP	---	E-MgATP	1.05	---	---	0.10	1.1	4.1
MgATP	+0.5mM CaATP	E-CaATP	---	---	0.015	6.67	0.9	6.5
---	---	E-Ca	---	0.02	---	5.0	---	6.4
MgATP	+2.26mM ATP^{4-}	E-ATP^{4-}	---	---	0.27	0.37	1.49	4.8
MgATP	+2.2mM Mg^{2+}	E-Mg	---	4.7	---	0.021	1.1	3.2
MnATP	+5.5mM ATP^{4-}	E-Mn	---	3.8	---	0.026	---	3.3

Equilibrium between Associated and Ionized Species of ATP in the Presence of Mg^{2+}, Mn^{2+} or Ca^{2+}

Km(MnATP)=0.065mM versus Km(MgATP)=1.05mM, correspond to 16 times lower affinity of the basal enzyme for the substrate MgATP than for MnATP. In table, dissociation constants show that an ionic equilibrium order Kd(Mg^{2+})=4.7mM>Kd(Mn^{2+})=3.8, indicating that Mg^{2+} can only at higher concentrations than Mn^{2+} displace the latter from the divalent metal activatory site.

The activity order MnATP>MgATP and both ATP^{4-} and CaATP act as inhibitors. Since Ki(ATP^{4-})=0.27 and Ki(CaATP)=0.015mM the competitive affinity order CaATP>MnATP>ATP^{4-}>MgATP.

At lower concentration of excess ATP^{4-} the greater Ka of the nucleotide for either Mn^{2+} or Mg^{2+} than for Ca^{2+} could maintain the enzyme active.

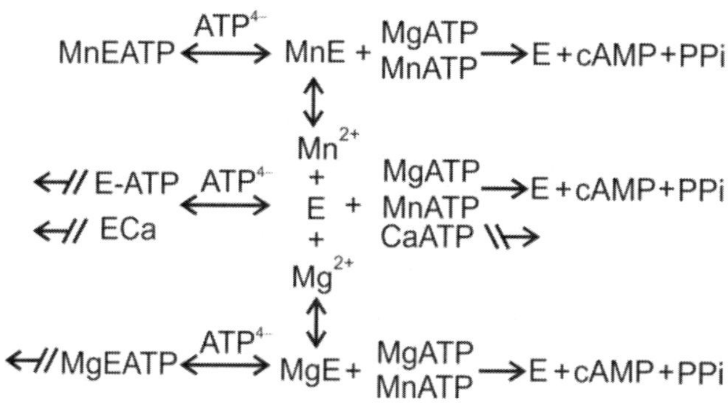

Scheme 2. Equilibriums for the RARE BiBi kinetic mechanism of the basal AC. Study of initial velocity patterns gave evidence that at the active site the substrates MgATP and MnATP by group transfer exchange PPi between ATP^{4-} and a pyrophosphate (PPi) accepting group, releasing cyclic adenosine monophosphate (cAMP) and PPi. Mn^{2+} substitutes for Mg^{2+} for substrate formation and as an enzyme modulator. The catalytic rates and dissociation constants for Mn^{2+} and MnATP differ from Mg^{2+} and MgATP with active site binding by inhibitor CaATP and ATP^{4-} resulting in dead-end inhibition (\\↦).

At higher concentration of excess ATP^{4-} the greater affinity (Ki=0.01) of the enzyme for CaATP, could predominate reducing the availability of cAMP for sustainable memory persistence. Hence, eradication of particular memory, confer to the system plasticity to re-conform synaptic connectivity for acquisition of a new memory or reinterpretation of the preceding one.

Mg^{2+} can easily displace Ca^{2+} from ATP and reduce the concentration of CaATP, because the association constant: Ka(MgATP)=20×10^3 M^{-1} is greater than Ka (CaATP)=9×10^3 M^{-1} [164, 165].

The mutual competitive inhibition of ATP^{4-} and CaATP on the substrates, MgATP and MnATP, and a constant cooperative value of about one indicate that the ionic and divalent metal complexes of ATP, bind to the same site on basal adenylyl cyclase.

A requirement of cAMP leading to increment neuronal connectivity implies that the level of AC activity may play a partial role of neuronal plasticity [166, 167]. This process results from a perdurable lack of cAMP to drive the creation and/or reinforcement required to maintain synaptic connectivity. The requirement to control cAMP levels indicates the applicability of AC kinetics for modeling the versatility of AC responsiveness. Enzyme plasticity is implied in the scheme 2, showing a random BiBi sequential group transfer mechanism (RARE BiBi) with dead-end inhibition, since the binding at the active site of CaATP or ATP^{4-} results in dead-end inhibitions.

In the intracellular fluid at 1mM [Mg^{2+}] is considerable higher than 1μM Ca^{2+} [168]. The ratio is practically constant, except for certain pathological states. However, the equilibrium dynamics of divalent metal existing as Mg^{2+} and MgATP, as well as Mn^{2+} and MnATP fluctuate. The latter, due to accumulation of endogenous metabolite intermediates like ATP^{4-}, amino acids, dicarboxylic acids, and organic phosphates, which can complex Mg^{2+} or Mn^{2+}. The cytoplasmic concentration of Ca^{2+} is in a dynamic state by the same metabolic variables. Additionally, 1mM extracellular and 10mM mitochondrial concentration of Ca^{2+} allow import and export of Ca^{2+} across structural compartmentalization [169, 170].

A strong restriction to the adenylyl cyclase activation, became evident from $S_{0.5}(ATP^{4-})$ of about 0.1mM. Hence, any decrease in the rate of ATP^{4-} consumption by the Na^+ gate activity and others functions, should turn off the adenylyl cyclase at both the neurotransmitter and basal levels. It is suggested that a matching the rate of ATP production and consumption should be optimal because the brain lacks storage capability. ATP production could overcome its consumption rate, and results in endogenous accumulation ATP^{4-}.

The nucleotide could tend to chelate the divalent metal (Mg^{2+}), which could be structurally present for binding receptors and the enzyme to the membrane [10]. Hence, neurons may need to adjust the rate of endogenous aerobic glycolysis to prevent ATP^{4-} accumulation, to overcome the stability of receptors binding into the membrane.

The RARE BiBi model predicts a complex response of AC at various total concentrations of $MgCl_2$ and $MnCl_2$ and the change in the Hill cooperativity for the $MgCl_2$ saturation curve from a basal value of 2.6 to 4.0 in the presence of noradrenaline (NA). This change in "n" may indicate that excess Mg^{2+} is obligatorily required for NA to stimulate the brain enzyme or the adrenaline-dependent stimulation of adenylyl cyclase in fat tissue.

The RARE BiBi mechanism also accounts for $CaCl_2$ inhibition of the enzyme, by the equilibrium concentration of CaATP, a competitor of the active site. Thus, CaATP is a dead-end inhibitor, the same as ATP^{4-}.

$$\frac{v}{Vm} = \frac{\frac{k_o}{k} + \frac{[Mg^{2+}]}{\alpha K_{di}}}{\left(1+\frac{[Mg^{2+}]}{K_{di}}\right)\left(1+\frac{[ATP^{4-}]}{K_{i1}}+\frac{[CaATP]}{K_{i2}}\right)\frac{K_m}{[MgATP]}+\left(1+\frac{[Mg^{2+}]}{\alpha K_{di}}\right)}$$

The formula was able to match without significant deviation the experimental vs computational saturation curves, obtained for the kinetic response of AC [165]. Computation based to the equilibrium concentration of different ionic species, according to the total [$MgCl_2$] or [$MnCl_2$] vs [NaATP] and the corresponding constants of inhibition (Ki) association (Ka) and dissociation (K_{di}) values [171, 172].

Chapter 6

The Neuronal-Astrocyte-Capillary Coupled System Role in Adrenergic vs Glutaminergic Neurotransmission

The activated by GTP-Gα subunit couple to activate adenylyl cyclase [ATP pyrophosphate-lyase (cyclizing) (EC 4.6.1.1)] (AC) [173] which converts MgATP into cyclic adenosine monophosphate (cAMP).

Adrenaline and noradrenaline (NA) bind to the cell membrane subtypes of GPCR for differential functions. Catecholamines released from the adrenal medulla, have effects mostly in peripheral tissues and not directly in brain, due to the lack of a catecholamine transport mechanism for the crossing of the blood-brain barrier (BBB).

The activatory mechanism of GPCRs, involves a guanine diphosphate (GDP) binding to an inactive Gα subunit, which became active when binds to a guanine triphosphate (GTP) to release GDP.

The Gα-ATP complex became released from 7TM-βγ complex to activate adenylyl cyclase. The Gα subunit and the βγ complex could each activate a variety of effector molecules.

The GPCR with Gα subunits of the four major classes (Gs, Gi, Gq, and $G_{12/13}$) provides for a very large number and diversity of the members within the 7TM family.

The 7TM β-adrenergic receptors have been subclassified as β_1, β_2, and β_3 receptors. These receptors have greater affinity for isoproterenol than for adrenaline or NA. The binding of adrenaline to the 7TM β-adrenergic receptors allows coupling with the Gs protein.

The presynaptic α-receptors are mostly involved in negative feedback loops to regulate the release of NA, but also play a role in blood pressure homeostasis.

The α_2 adrenergic receptor (or adrenoceptor) (α_2AR) is a hybrid associated with the inhibitory Gi heterotrimeric G-protein (Gi) in a transmembrane sequence, resembling the β_2 subtype (βAR) that in the muscarinic and acetyl choline (Ach) receptors inhibits AC.

If cAMP is not inactivated by conversion into 5'-AMP by the action of the enzyme cyclic nucleotide phosphodiesterase (PDE) acts as a second messenger in the cytosol for activation of protein kinase-A (PKA).

Adenylyl Cyclase

The adenylyl cyclase (AC) family is composed of ten isoforms: nine of them are membrane-bound (AC1–AC9), while one of them is soluble (AC10). The domain structure of AC1–AC9 is characterized by having two regions where there are six transmembrane regions. The large cytoplasmic domains C1 and C2, which contain the catalytic region, form a heterodimer and co-operate with each other to convert ATP into cyclic AMP.

The presynaptically discharged noradrenaline (NA) acts postsynaptically on the Mg^{2+} sympathetic receptor or on the Ca^{2+}-dependent α_2-receptor for a separated AC responsiveness to an activator free Mg^{2+} or inhibitor Ca^{2+}. The Mg^{2+} incremented cyclic adenosine monophosphate (cAMP) became a ligand of the cellular transcription factor, cAMP response element-binding protein (CREB). This one is generally accepted to participate in the plasticity of the neuronal circuits involved in long-term memory. Thus, allows an integrative role to the brain adrenergic system, synchronizing the glutamate-mediated neurotransmission. Additionally, AC variable responsiveness to NA also fits with the effect of Ca^{2+} vs Mg^{2+} in locally adjusting the glucose permeability of each capillary-neuronal BBB.

Cyclic AMP Signaling Pathway

Cyclic adenosine monophosphate or 3'-5'-cyclic adenosine monophosphate (cAMP) is a ubiquitous second messenger that regulates a multitude of cellular responses. cAMP formation usually depends upon the activation of G-protein coupled receptors (GPCRs) that use heterotrimeric G-proteins to activate the amplifier adenylyl cyclase (AC), which is a large family of isoforms that differ considerably in both their cellular distribution and the way they are activated. There are a number of cyclic AMP signaling effectors such as protein kinase-A (PKA), the exchange proteins activated by cyclic AMP (EPACs) that activate the small GTP-binding protein Rap1 and the cyclic nucleotide-gated channels (CNGCs). These various effectors are then responsible for carrying out the cyclic AMP signaling functions that include control of metabolism, gene transcription and ion channel activity. In many cases, these functions are modulatory in that cyclic AMP often acts to adjust the activity of other signaling pathways and thus has a central role to play in the cross-talk between signaling pathways. This modulatory function is particularly evident in the case of Ca^{2+} signaling in both neuronal and muscle cells. Many of the functions of cyclic AMP depend upon the precise location of PKA relative to both its upstream and downstream effectors. A family of A-kinase-anchoring proteins (AKAPs) determines this cellular localization of PKA as well as a number of other signaling components. The off-reactions are responsible for removing cyclic AMP by either hydrolysis or by efflux from the cell.

Receptors

AMPA receptor (α-amino-3-hydroxy-5-methyl-4-isoxazolepropionic acid) a GluA2 (GluR2) tetramer is a non-*N*-methyl-D-aspartate (NMDA)-type ionotropic transmembrane receptor was the first glutamate receptor-ion channel to be crystallized, activated by [3H]AMPA analog of glutamic acid. AMPARs that mediate fast synaptic transmission and it is the most commonly found receptor in the central nervous system (CNS) [174].

The γ-aminobutyric acid (GABA) receptors are a class of receptors that respond to the neurotransmitter GABA, the chief inhibitory neurotransmitter in the vertebrate central nervous system.

$GABA_{(A)}$ an ionotropic receptor is a ligand-gated ion channels, activation, conducts Cl^- through its pore, resulting in hyperpolarization of the neuron. Its inhibitory effect decreases transmission of successful action potential at pH sensitive some subunits are activated and some inhibited by extracellular protons [175].

The reversal potential of the $GABA_{(A)}$-mediated IPSP in normal solution is −70 mV.

The protein contains different allosteric binding sites for the nonbenzodiazepines, ethanol, barbiturates, benzodiazepines, [176] neuro-active steroids, inhaled anaesthetics and picrotoxin.

$GABA_{(B)}$ receptors are metabotropic transmembrane receptors for GABA coupled to G-protein and linked to K^+ channels [177] are found in the central as well as in the autonomic division of the peripheral nervous system. $[K^+]$ hyperpolarize the cell at the end of an action potential. The reversal potential $GABA_{(B)}$-mediated IPSP (inhibitory postsynaptic potential) is −100 mV, more hyperpolarized than the $GABA_{(A)}$ IPSP.

Acid-sensing ion channels (ASICs) near in the post-synaptic membrane are proteins that could respond rapidly to acid transients, affecting extracellular acidosis, this inhibits NMDA receptors [178, 179, 180] and AMPA receptors [181].

Dopamine and Serotonin Pathways

Receptors could act through diverse G-protein coupled and G-protein independent mechanisms that trigger downstream intracellular signal transduction events involving the cyclic adenosine monophosphate (cAMP)/ protein kinase-A (PKA), PI-3Kinase/Akt, phospholipase A2 (PLA2), and phospholipase-C (PLC) pathways. These pathways in turn regulate various functions including synthesis, transport and degradation of dopamine and serotonin as well as the transcriptional regulation of key genes linked to multiple neuro-pathological conditions.

Dopamine and serotonin are 2 of the major neurotransmitter systems in the mammalian nervous system. Dopamine affects brain processes that control both motor and emotional behavior and plays a role in the brains reward mechanism. Serotonin is critical in temperature regulation, sensory perception, locomotion, sleep, and psychosis.

Magnesium affects the function of the serotonin receptor nitric oxide synthesis and release, which are necessary for mood stabilization, cognitive function, and brain health.

Neurotransmitters, Mg^{2+} and Ca^{2+} Modulation of Receptors and Ion Channels

All levels of the central nervous system (CNS) are innervated by LC-noradrenaline (NA) axons with providing NA innervation at the isocortex, allocortex and cerebellar cortex [32,33, 182, 183], and additionally, caudally innervate the medulla and spinal cord [184, 185]. A single NA-neuron through its branching could spread with terminals over most of the entire cerebral cortex.

Divergence of axon collaterals occurs implicating non-synaptic effects by diffusion on neighboring cells, acting as a neuromodulator within circumscribe microenvironments but widespread allowing NA to reach many brain regions [186, 187].

The central NA system paracrine effects appear to be related to the marked divergence sustained by the morphology of the axon terminals. The collaterals possess varicosities or boutons, and spikes across the axon release NA from multiple serial sites, in addition to the one present at the fibre endings.

The varicosities along LC-NA axons possess 0.5 μm beaded varicosities that differentiate from the larger diameter 1 to 3 μm medullary A1 and A2.

Cocaine and some antidepressants interfere with the reuptake process of norepinephrine. This prolongs its synaptic effects.

α_1 antagonists leave the presynaptic α_2 receptor intact and functioning. This will reduce the total norepinephrine released which will prevent the tachycardia.

At low doses, dopamine can act on β_1 receptors; at higher doses on α_1 receptors. It also causes the release of norepinephrine from the nerve terminals, which will also stimulate β_1 and $\alpha 1$ receptors. This will ultimately act to increase renal blood flow, increase cardiac output, and increase total peripheral resistance.

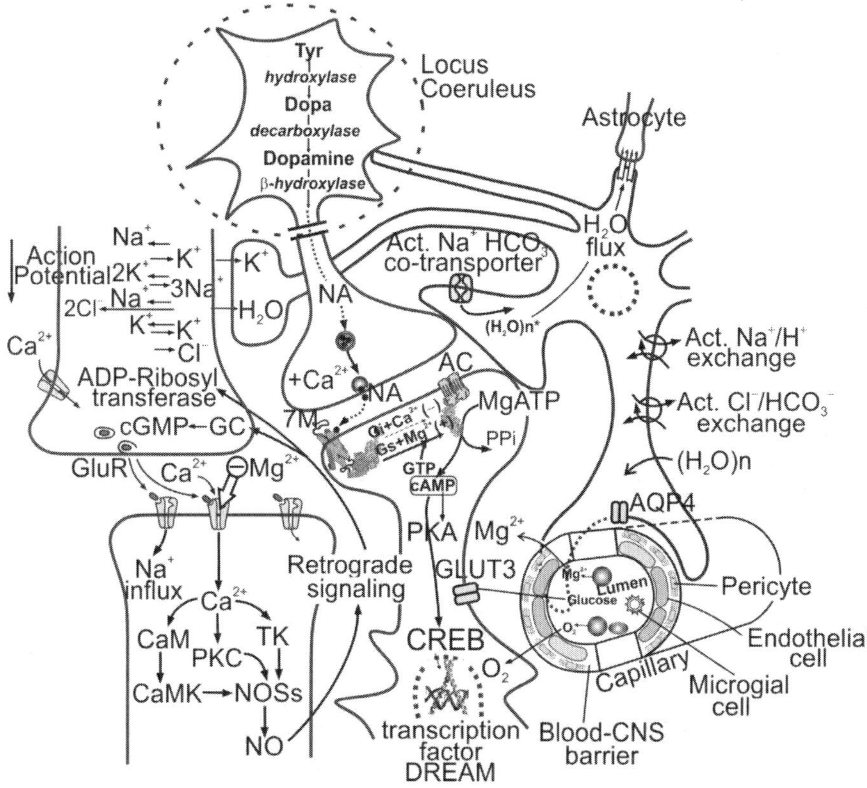

Figure 25. Locus coeruleus axons presynaptically release noradrenaline (NA) for activation of a postsynaptic adenylyl cyclase. G-protein stimulated by α-receptor (Gαs) in a Mg^{2+}-mediated coupling stimulate the postsynaptic NA-AC. The G-protein in Ca^{2+}-mediated coupling with the α-receptor inhibit (Gαi) the postsynaptic basal AC. Glutamate receptors (NMDA) are inhibited by Mg^{2+}, the ion activating AC. Calmodulin (CaM), protein kinase C (PKC), tyrosine kinase (TK), nitric oxide (NO), nitric oxide synthase (NOSs), cyclic GMP (cGMP), guanylyl cyclase (GC), cyclic adenosine monophosphate (cAMP) response element-binding (CREB), glucose transporter-3 (GLUT3), NA receptors with Ca^{2+} + heterotrimeric-G-protein-α-inhibitory (Gαi) result in decrease of cAMP. Mg^{2+} + heterotrimeric-G-protein-α-stimulatory (Gαs) increases cAMP production [188]. Mg^{2+} may bind to the transcription factor DREAM (calsenilin/KChIP3) and alter transcription.

Acetylcholine release onto preganglionic fibers causes depolarization, calcium influx, and secretion of epinephrine. Glucocorticoid levels (elevated in prolonged stress) will induce the activity of PNMT (phenylethanolamine N-methyl transferase), resulting in increased synthesis of adrenaline.

The α and β antagonists are used to block to catecholamine excessive effects on the receptors in preparation for surgery.

The NA axon activity could spread to glial (astrocyte) and the microglial release of specific cytokines [189, 190] and affecting permeability of the capillary, including modulation of the Mg^{2+} and O_2, releases across the blood-brain barrier [191].

Increment of Ca^{2+} in cerebrospinal fluid (CSF) at the synaptic level could activate the glutaminergic neuron, and inhibits the noradrenergic neurons allowing the formation of the G-protein-α-inhibitory subunit complex (Gαi-receptor). The effect inhibiting AC could decrease the generation of cyclic adenosine monophosphate (cAMP) bellow basal levels.

Catecholamines are degraded by MAO (monoamineoxidase) and COMT (catecholamine-O-methyltransferase). The resulting metabolite is 3-methoxy-4-hydroxy-mandelic acid (aka VMA).

The cAMP increment by AC could initiate a voltage-gated nerve impulse because the tendency of a more reactive charge expressed by the Mg^{2+} partially deprived of hydric shell to displace Na^+ from the outside into the inside of the membrane pore/channels. Neurotransmitters glutamate and ATP are agonists of receptor-operated channels (ROCs) present in the plasma membrane for uptake of external Ca^{2+}. The latter, as an intracellular second messenger role involves sensors that act through effectors Ca^{2+}-sensitive K^+ channels, Ca^{2+}-sensitive Cl^- channels (CLCAs). The buffer parvalbumin (PV) do not respond to a rapid Ca^{2+} signals onset, but soaks up Ca^{2+} for relaxation. The increased volume of mitochondria couples to modulate the PV ability to remove Ca^{2+} during the recovery/reuptake phase.

Downstream regulatory element antagonist modulator (DREAM [192] like calsenilin and KChIP3 [193]) is structure of 29-kDa EF-hand Ca^{2+}-binding protein functions as a transcriptional repressor by controlling prodynorphin expression and regulates pain-transmission. It translocates into the nucleus during calcium signaling for c-*fos* genes and blocks transcription by binding to downstream regulatory element (DRE) [194, 195, 196]. The cAMP has effect on gene expression in the nucleus of the cell and in the processes of growth and development.

Cerebrum cells, lack of storage pathways are nutritionally dependent of a continuous uptake of glucose and oxygenation to support aerobic glycolysis. Under starvation ketone bodies can only partially substitute for glucose. The need for signal transduction between tissues involved, required for matching the rates of O_2 and glucose consumption. This capability allows cerebral areas to increase by 10 times the rate of aerobic glycolysis without overcoming the homeostatic controls of glucose level in blood.

The Neuronal-Capillary-Astrocyte Network for Mg^{2+}-Activated Adrenergic vs Ca^{2+}-Activated Glutaminergic Modulation of Neurotransmission

Astrocytes are electrical not excitable cells which do not respond directly to glutamatergic signaling evoked by sensory stimulation.

In the brain, the pulsation pressures on cerebrospinal fluid (CSF) flux between para-arterial spaces and paravenous pathways drive CSF influx and directionality [197] along astrocytes.

Astrocytes modulate neural network activity by a Ca^{2+}-dependent transient uptake of extracellular K^+, which led to neuronal hyperpolarization and a decrease of the frequency baseline, improving the signal-to-noise and fidelity on postsynaptic transmission [198]. The evolutionary increment of brain capabilities were supported by the role of astrocytes to keep a homeostatic temperature. This cell drives a rapid flux of H-bond depleted water cluster out of the system.

Ion channels transverse the lipid bilayer to orchestrate at the cell membrane electric signals for function of central nervous system, peripheral nerve, and both skeletal and cardiac muscle [199, 200, 201].

Cloning and expression of genes that encode channel proteins has made the study of expressed proteins at the level of single channel molecules possible. It also reveals the pathogenesis of numerous disorders [202].

The skeletal muscle type 1 Na^+ channel promoter E-box interacts with an upstream repressor element with release of repression by myogenin [203, 204].

The interdependent of adrenergic receptors with other neuronal transmitter pathway constitute a network that potentiates and endows the brain with a potent working memory, but extremely sensitive to stress. Acute-panic could be associated to a lack of a promptly reuptake of noradrenaline (NA), allowing a symptomatology of anxiety followed by depression, due to the exposure of the NA-AC complex to the time-dependent inactivatory effects of NA on the AC enzyme. The figure 25 and 26 illustrates that when Mg^{2+} is released acts inhibiting the postsynaptic *N*-methyl-D-aspartate (NMDA) receptors and the entrance of Ca^{2+} into the dendrites. The effect prevents the inhibition of postsynaptic adenylyl cyclase.

A stimulated cluster of neurons starts firing and simultaneously the diameter of the surrounding arterioles increase to deliver a larger volume of blood, decreasing stimulation lead the blood-vessels to contract [197, 205.

C-fibers the NMDA and neurokinin-1 (NK1) antagonist receptors induce a long-lasting cumulative depolarization and through second-messenger potentiate the NMDA regulation of excitability of ion channels action potential response.

Ca^{2+} increases the strength of attraction between the endothelial cells of the capillary BBB and Mg^{2+} relaxes the tightness of interaction, facilitating during deoxygenation its crossing by the Hb-released O_2 and Mg^{2+}.

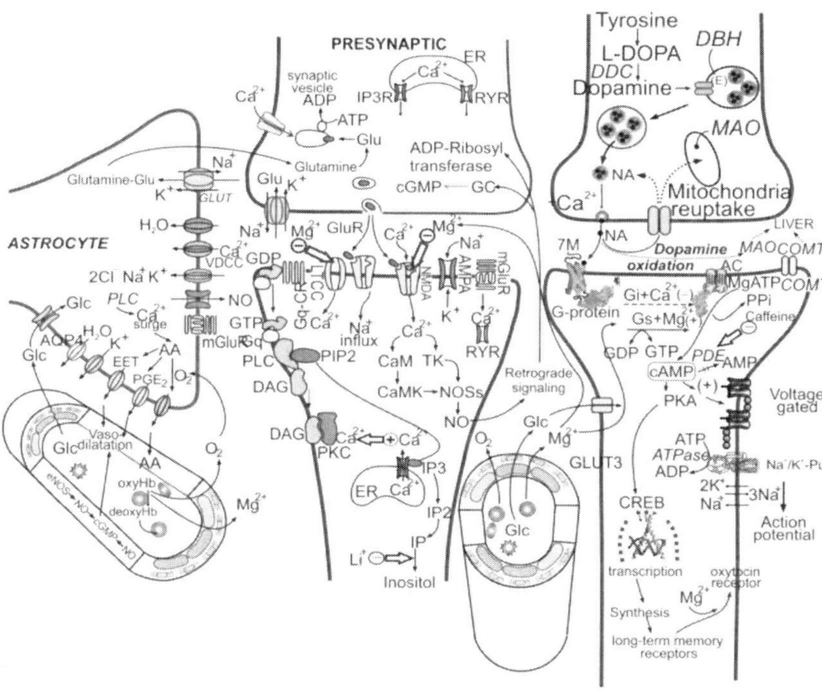

Figure 26. Ionic-metabolic integrative coupling on the neuronal astrocyte-red cell system. G-protein-coupled receptor (GPCRs) exchange its binding to GDP inactive for GTP active (GDP/GTP-exchange) for NA receptor coupling to AC. Neuron uptake and stoichiometric cycling reversing the glutamate synthase reaction supports release of glutamate (Glu), activating the postsynaptic N-methyl-D-aspartate (NMDA) receptor. At the astrocyte synaptically released GLU is co-transported with Na^{+} by the Na+/K+ ATPase consuming one ATP in the exchange for extracellular K^{+}.

The G-protein-coupled receptor (GPCRs) α, β, γ subunits dissociate into separate α and βγ activated complexes. Once the α-subunit hydrolyzes bound GTP to GDP, it re-associates with the βγ complex for ending receptor activation. The system has a requirement for free Mg^{2+} for AC conversion of MgATP or MnATP into endogenous cAMP. cAMP could be converted by phosphodiesterase (PDE) in AMP or could activate protein kinase-A (PKA) with subunits C2R2, by dissociating regulatory (R) from catalytic (C) units. cAMP-C acts by phosphorylating proteins and enzymes, changing these from inactive to active forms or vice versa. The WM relates to phospholipase C [172] and endogenous intracellular IP3-mediated Ca^{2+} level [172, 206, 207].

The joint release of Mg^{2+}, glucose and O_2 allows the matching of O_2 release to that of glucose consumption rates, during brain aerobic glycolysis.

The overall metabolic feedback by the *in common* dependence on free Mg^{2+} or Mn^{2+} of AC of brain, fat cells and the insulin receptor tyrosine kinase (IRTK) in liver, allows homeostatic network integration.

The main astrocyte role would be to manage a water circuit turnover of CSF to maintain thermic and larger aggregates of water into homeostasis state. Within the open neuronal-astrocyte network system the smaller aggregates of water are released into venous blood and their exclusion from the body allows to prevent entropy increase by the highly energy demanding neuronal processes.

The overall process requires preserving energy of activation and processing of information, separated by the neural lipid-protein membranes to proceed to its completion without reaching thermic homeostasis.

Thermic equilibrium is delayed in part by the astrocytes taking-out of system the H-bonds depleted water, which could be thereafter associated with air-water exhaled-out to drive heat/entropy to the outside as described for other open systems.

Aquaporins

Several aquaporins are found in the brain including aquaporin 1 (AQP1), AQP3, AQP4, AQP5, AQP8 and AQP9 with AQP1, AQP4 and AQP9 most heavily studied.

Aquaporin 4 (AQP4) is a protein which in humans is encoded by the AQP4 gene [208]. AQP4 belongs to the aquaporin family of integral membrane proteins that conduct water through the cell membrane. Whereas

AQP9 is found in the astrocyte cell body, AQP4 is abundant in perivascular astrocyte end feet and also where the astrocyte is in close apposition to neurons [205].

Choroid plexus exhibits AQP1 and AQP4, and endothelial cells of the neurovascular unit (NVU) configure conductances pathways [209].

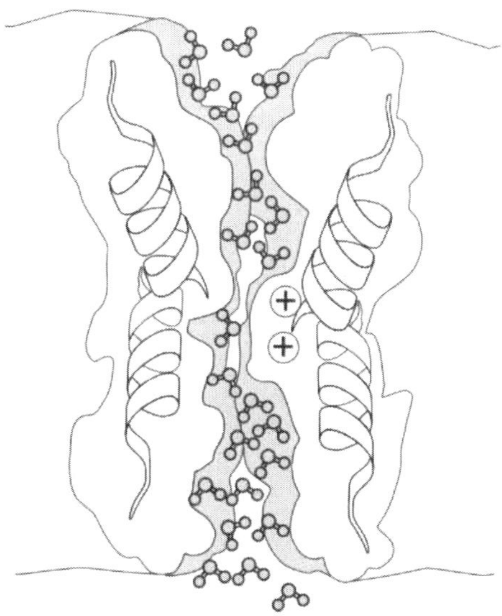

Figure 27. The gases do not move in aquaporin by a simple diffusion but AQP allows facilitated diffusion through which water, CO_2 and NH_3 moves through the endothelial cells of the NVU.

Astrocyte's aquaporin could be upregulated by a direct insult to the central nervous system and control of water distribution within the brain and show associated fluid dynamics dysfunctions [210].

The Coupling of Neurons, Glial and BBB Modulates Metabolism and the Rate of the TCA-Cycle

Figure 28 illustrates that entwined on astrocytes the capillaries enveloping the endothelial cells allow a Ca^{2+}-contracting and Mg^{2+}-distending the tightness, between pericytes to modulate access across the blood-brain-barrier (BBB).

BBB occupies the largest surface area delimiting Blood-CNS (central nerve system) interface other barriers: the blood-retinal, blood-nerve, blood-labyrinth, and endothelial of brain with the choroid plexus to form the blood-cerebrospinal fluid (CSF).

^{13}C NMR spectroscopy allows detecting resonances from ^{13}C, the only stable isotope of carbon having a magnetic moment. The natural abundance for ^{13}C is approximately 1.1% of the total carbon and its magnetogyric ratio is approximately one-fourth of that of the proton. The technic sensitivity can be improbed using ^{13}C-enriched substrates in specific position allow to follow in vitro and in vivo the activity of a large variety of metabolic pathways. These include glycolysis and the pentose phosphate pathway, glycogen synthesis and degradation, gluconeogenesis, the TCA cycle, ketogenesis, ureogenesis, and the Glu–Gln/GABA cycle in brain, among others [211, 212, 213, 214, 215]. This provides an in vivo evaluation of cerebral metabolic compartmentalization [216, 217].

In neonates the transport of glutamine from astrocytes to GABAergic neurons was lower. The literature reports that in astrocytes by using [1,2-^{13}C]glucose it could be shown that despite much lower pyruvate carboxylation, relatively more pyruvate was directed from glycolysis towards the Anaplerotic reactions than to pyruvate dehydrogenation [218, 219].

In addition to glutamate (Glu) actions on these receptors, astrocytes also uptake Glu through Glu-transport protein-mediated endocytosis and then astrocytes recycle the Glu back to neurons for further use by synthesis of glutamine (via glutamine synthetase, exclusively in astrocytes) with secondary release for neuronal uptake and repackaging of Glu. This recycling creates a Glu-glutamine cycle occurring between neurons and astrocytes [220]. The vesicular uptake of Glu requires Glu-transport protein and vacuolar-type H^+-ATPase, which couples ATP hydrolysis and Glu-transport [221] and which also make the interior of the vesicle acidic. Glutamine uptake into neurons increases during hypoglycemia since glutamine can also be oxidized for energy production in low glucose conditions [222].

The glucose uptake from plasma by neurons is mediated through the glucose transporter-3 (GLUT3). At the astrocyte level glucose became transported through GLUT1 and glycolysis supports a continuons conversion of lactate to pyruvate (Lac/Pyr) by lactate dehydrogenase isoenzymes (LDH5) in astrocyte and LDH1 in neuron. This redox shuttling releases Lac into extracellular surroundings for neuronal uptake as metabolic fuel. The astrocyte glutamine synthase (Glu synth.) consumes ATP to form glutamine (Gln).

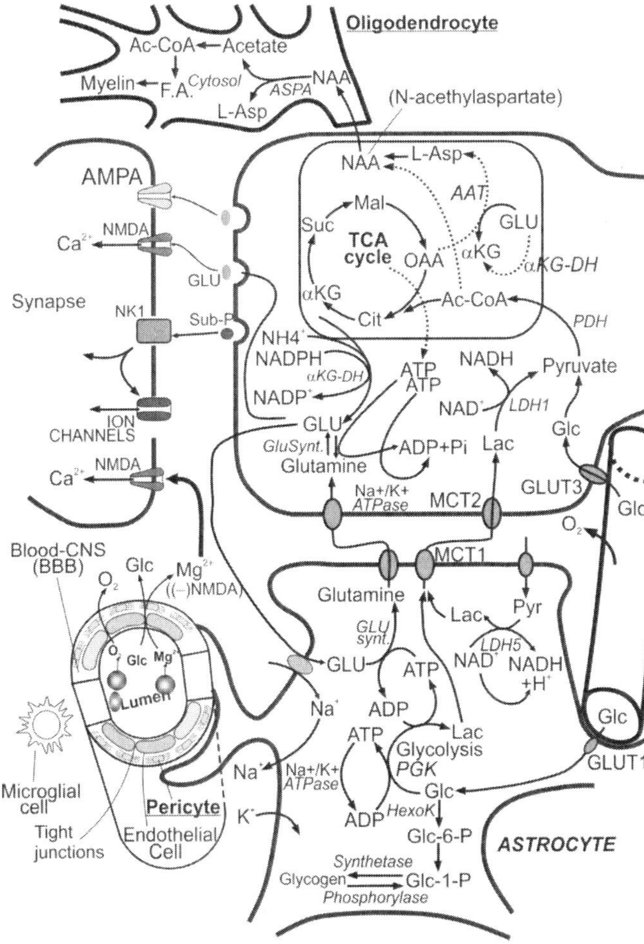

Figure 28. Neuron-astrocyte-BBB-(blood-brain-barrier) coupling on glutamatergic-(GLU)-neurotransmission. The neuronal mitochondria acetyl CoA and L-aspartate became the substrate to synthetizes N-acetylaspartate (NAA) a precursor with l-glutamate to form N-acetylaspartate-glutamate (NAAG), which at the astrocyte could be broken by carboxypeptide II, releasing L-glutamate and NAA. At the oligodendrocyte cells NAA by action of the enzyme aspartoacylase (ASPA) releases L-aspartate (L-Asp) and acetate.

Acetate, as acetyl CoA (Ac-CoA) became the substrate for fatty acid and support of myelin synthesis. In Canavan's disease has been observed that a defective N-acetylaspartate catabolism reduces brain acetate levels and myelin lipid synthesis.

Glycolysis, the TCA cycle and the pentose phosphate pathways break down glucose from carbohydrates into the metabolites necessary for energy production, and gluconeogenesis stores excess energy as glucose.

The ASPA gene instructions for producing the enzyme N-acetylaspartate amidohydrolase (aspartoacylase), in the brain, this enzyme breaks down [223].

N-acyl-L-aspartate (NAA) + H_2O Acetate + L-aspartate

The connection between ASPA deficiency and the failure of proper central nervous system (CNS) development is not clear.

Chapter 7

The Dynamic Regulation of the Brain Barrier Permeability by the Neurovascular Unit (NVU)

The vascular wall consists in the endothelial cells connected by tight junctions, resting on the basal lamina, and the choroid plexus (which forms the blood-cerebrospinal fluid (CSF) barrier). It includes pericytes, smooth muscle cells and astrocytic end feet covering about 98% plus occasional neuronal terminals. The lymphocytes and monocytes circulating blood cells adhere and roll along the vascular lumen and perform surveillance of neural signaling and cellular activity [224]. The astrocytes extend processes surrounding synapses to link neuronal activity with the oxygen and nutrient supply.

Neurovascular unit (NVU) [225, 226, 227, 228] consisting of 10-15% of all proteins detected and localized in the vascular endothelial cell monolayer with the role of transporter. These, generally function as efflux and coupled sodium- and chloride-dependent GABA-transporter, noradrenaline-transporters (NeT) family proteins, excitatory amino acid-transporters (eAATs) and the ATP-binding cassette (ABC) transporters. These utilize the energy of ATP binding and hydrolysis for translocations of substrates across membranes and for the non-transport-related translations of RNA and DNA repair processes [229, 230]. All these proteins integrate a superfamily present in all phyla from prokaryotes to humans [231, 232].

Astrocytes and pericytes possess their own complement of transporters, channels, receptors and signaling mechanisms with which they coordinate the role of the NVU in supporting nervous system function [233].

Cells of the NVU form a complex and fine-tuned transport system that balances the influx of nutrients and the efflux of wastes, toxins and drugs to maintain central nervous system (CNS) homeostasis. NVU is a dynamic structure consisting of endothelial cells that cerebral ischemia affects to lead to brain edema. The ischemic insult treatment induces shedding of low-density lipoprotein receptor-related protein (LRP)'s ectodomain from perivascular astrocytes into the basement membrane.

The most important roles of the NVU is to limit xenobiotic, including drugs for CNS treatment from entering the brain through the participation of two components in the brain capillary endothelium, the ATP-driven membrane transporters known as *efflux transporters*. The endothelial cells of the NVU are highly polarized with different integral membrane proteins at the luminal and abluminal surfaces. Tight junctions *seal* the inter-cells space at the monolayer of the endothelial capillary in the presence of Ca^{2+}, but respond to Mg^{2+} distending their contacts.

Endothelial cells-tight junctions when broken became surrounded by microglial cells covering the damage areas.

The ion transporter proteins in the cells of the NVU maintain fluid balance in the brain, water and electrolyte movement, through a family of HCO_3 transporters, the electrogenic sodium bicarbonate cotransporter (NBCe) and electroneutral sodium bicarbonate cotransporter 3 (NBCn) families, and electroneutral sodium-driven chloride/bicarbonate exchanger 1 (NDCBe).

The expression of these transporters vary with brain region and cell type with prominent expression of both NBC and NDCBe transporters reported for neurons and choroid plexus, and little or no expression in astrocytes.

Numerous factors regulate the barrier permeability of the NVU, including modulation of membrane transporters and transcytotic vesicles, and modulation of trans-cellular permeability.

Parturition and nurture events link through the mother oxytocin, produced by the hypothalamus and stored and secreted by the posterior pituitary gland, create the mutual bonding and recognition learning of her to the new child.

Thyroid hormones concentrated in nerve terminal and released, may have non-genomic mechanisms of action as putative neurotransmitters or neuromodulators influence the binding and function of the γ-aminobutyric acid-A ($GABA_{(A)}$) receptor, chief inhibitory neurotransmitter in the mammalian CNS.

Thermodynamics of ATPases on Electrogenic Membrane Systems

Chloroplast fragments have a requirement for uncouplers, in other to reverse the light dependent ADP-phosphorylation, which otherwise does not have any detectable ATPase [125, 126].

The catalytic restriction of CF1 to maintain only a forward direction had been assumed not to be compatible with the principle of microscopic reversibility. The latter predicts by analogy with an open door that molecules could transit in both senses.

It is widely known how macroscopic rotation doors by revolving around an axis allow only transit by a unidirectional revolving sense.

At the molecular level, the studies of chloroplast factor I (CF1) were shown to require heating to evidence Ca^{2+}-dependent ATPase activity in the absence of light. CF1 binding to the chloroplast membrane [10] required Mg^{2+} or Ca^{2+}, but heat-activation or trypsin-treatment, prevented a recombination.

Membrane extraction with solvents resulted in the separation of a soluble and one insoluble fraction. Combination and time-dependent incubation recovered the membrane allotopic property of increasing the cold stability of CF1 protein. Therefore, membrane vs enzyme location, allows the CF1-ATPase to rotate over an axis, and differentiate exterior vs interior space. CF1-ATPase has two active catalytic sites of different affinities for S and P.

In an open system CF1-ATPase has two active catalytic sites of different affinities for S and P. The single rotational sense allows the membrane to time-differentiate the protein rotational-conformation required for ligand ions and water uptake from the outside to their releasing inside the cell.

A single directional sense is illustrated in figure 29 describing the Na^+/K^+-ATPase catalytic conformational rotation in electrogenic membranes from steps 1 to 4. Rotation allows a time differential for the physic-chemical events required for ligand ions and water uptake from the outside to their releasing inside the cell.

The parameters for the thermodynamic reaction progress (figure 7) in the steps sequence 1 to 4, allow the enzyme rotation over its axis to link events in asymmetric thermodynamics. This configures the conditions allowing overcoming thermic dynamics by the dissipative potential of H-bond strutured H_2O-protein system. These are capable to release H-bond energy to bypass microscopic reversibility and support vectorial kinetics. The breaking of transitive H-bonds in the hydration shells of protein-ions requires 10^{-12}s, allowing the less hydrated ion forms to participate in the conformation of action potentials.

Prigogine described without giving biological models, the thermodynamic role of a dissipative structural potential. The structure, in this case, could be the enzyme potential to change its hydration states. The caloric increment by the enzyme-membrane conformational changes implies an increase in entropy.

Similitude with a car, which functions as a gasoline vs. exhaust open system, implies heating of its motor, and entropy increase that has to be dissipated by cooling. The increase in the number of H_2O molecules in the hydration spheres of proteins, by turnover, is decreased, and lead to the loss of H-bonds. The thermodynamic of these bonds could be linked to the one of surrounding water. In this one a decreasing of cluster sizes allows the uptake of entropy/heat by CSF or blood to the outside of the system.

Tendency to return to ligand state is decreased by the capture of water molecules from the hydrated shells Na^+ and K^+. The overall equilibrium tendency depends at the last instance from the loss of H-bonds within water cluster, which explains the water-protein dependency for functional turnover.

Step 1: Illustrates the initial/final state of coupling of the ion pump-ligands as a function of the overall cytoplasmic domain nucleotide-binding (N) and a phosphorylation (P), interacted through three rotational states for the enzyme cycle E_1 (catalytic) and E_2 (modulatory) of ATPase-ligand interactions. At sites that are 50 Å apart are located the phosphorylation and the ion-binding sites. The cyclic adenosine monophosphate (cAMP) interacts for control of the voltage gated channel opening to the inside for transfer of $3Na^+$ from the outside, initiating of an action potential which producing a change in polarity across the axon membrane. In the cytoplasm side or internal E_1, the protein factor-1 (F_1) binds ATP to the active site in the β_E-subunit of the ATPase.

Step 2: Autophosphorylation by the ATPase of an aspartate (Asp) residue lead to an E_1 forms in a high-energy E_1P.ADP-complex for binding of the $3Na^+$ cation for their release to outside and $2K^+$ occupying their binding sites. The ATPase rotates anticlockwise 120° with regard γ-subunit. In the external side the E_2 state of the transporter has binding sites of decreased affinity for the Na^+ cation (Albers, 1967; Post et al., 1969).

Step 3: The induced conformational change transports Na^+ ions to the outer cell site of the transporter in a transition: E_2P to $E_2 \rightarrow E_1 + Pi$ couple to the release of $3Na^+$ into the outside.

Step 4: The autodephosphorylation of Asp starts another conformational change, and the A-domain rotates (120°) into the emerging gap of the phosphoenzyme. The geometry change decreases the affinity for the bound K^+ ion, releasing it into the inside. The Pi is released in the connection between the N- and P- domains, the P- moves up and down relative to the membrane.

Exclusion from hydrophobic regions of the protein-lipid electrogenic membrane and the proteins themselves allow one-sided distribution of ligand water sites.

Kosmotropic versus Chaotropic Tendencies Adjust the Size of the Hydration Shells of Na^+ and K^+ Ions for Fitting into the Gate-Dynamics of the Na^+-Ion-Pump

At 37°C the membrane equilibrium potential= −98mV, corresponding to $[Na^+]_{in}$=14mM and $[Na^+]_{out}$=143mM, potential= +62mV. For $[K^+]_{in}$=157mM and $[K^+]_{out}$=4mM, potential= −98mV. The enzyme is depicted as viewer from the foot of the central stalk. Only the catalytic β-subunits and the γ-subunits are shown. An action potential opens voltage-sensitive calcium channels in the presynaptic membrane, the influx of calcium lead noradrenaline (NA)-vesicles to release NA into the presynaptic cleft. In the arteriole close to the gap, oxyHb conversion to deoxyHb releases Mg^{2+} activating the NA ligante state of the postsynaptic G-protein α_S-receptor linked to adenylyl cyclase.

Total cell Mg^{2+} would be in the range of 10mmol/L if it were all free in the cytosol. Hence, as cytosolic Mg^{2+} levels are reported to be 0.5 to 1 mmol/L= 90% to 95% of the cell magnesium is bound or sequestered. Ionized Mg^{2+} levels in mitochondria are reported to be in the 0.5 to 1 mmol/L range and Mg^{2+} levels in SR are reported to be 1 mmol/L[5]; therefore, most of the magnesium in the cell is bound, regardless of the precise location. The electrochemical gradient across the membrane with regard to extracellular $[Mg^{2+}]$= 1 mmol/L and cytosolic $[Mg^{2+}]$=188 mmol/L involves a membrane potential of -70 mV, Mg^{2+} permeability is low and therefore Mg^{2+}-transporters do not require a high Vmax.

An action potential induces a depolarization that opens the presynaptic channel resulting in vesicular exocytosis. The neurotransmitter binds to postsynaptic receptors. The ionotropic opens the associated ionic channel and produces an ionic flux changing the membrane potential of the postsynaptic neuronal. Depolarizing membrane potential, in the glutamate (Glu) receptors, increases neuronal excitability and result in action potentials.

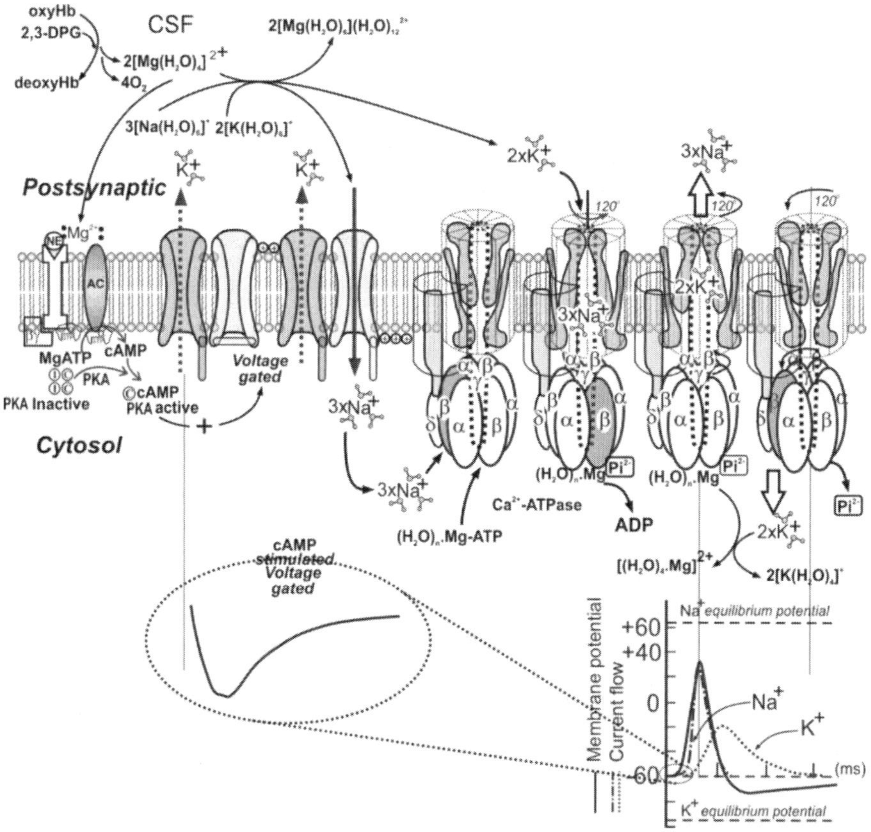

Figure 29. Mg-driven Interaction of ion-hydration shells for the coupling of Na^+/K^+-ATPase and AC. The Na^+-in channel and the K^+-out channel open as a function of changing potential ΔV. MgATP breakdown by the ion pump-ATPase is required for releasing Mg^{2+}, which when in excess of substrate became activatory of basal and NE stimulated AC. The differential strength between the tendencies of ions to complete their hydration shells allows directional reactivity. Moreover, the interactions of water depend of the different and variable stabilities for structuring H-bonds. Exclusion from hydrophobic regions of the protein-lipid electrogenic membrane and the proteins themselves allow one-sided distribution of ligand water sites.

Hyper-polarization in the γ-aminobutyric acid (GABA) receptors decrease neuronal excitability and reduce tendency to generate an action potential. There is 0.3 to about 5 millisecond synaptic action delay, between the arrival of the presynaptic signal and postsynaptic response. In the synapsis, the protein domains cross the neuronal membranes exhibiting an extracellular neurotransmitters recognition side.

The ionotropic receptors by association of subunits form the ionic channel that by binding the excitatory neurotransmitters: acetylcholine and Glu undergo conformational changes, which in a few milliseconds opens a channel to the cation Na^+, K^+ and Ca^{2+} or anion Cl^-. The inhibitory neurotransmitter $GABA_{(A)}$ and glycine act by hyper-polarization of the resting membrane. The presence of a receptor generates differential functions with an amply capability to regulation.

The Na+/K+-ATPase of the plasma membrane [123] is couple to a cycle in which each molecule of ATP becomes converted to ADP and Pi, moving $2K^+$ ions inward and 3 Na^+ ions outwards. This maintain a membrane potential of −50 to −70 mV (inside negative relative to outside) as result of supporting low Na^+ and high K^+ concentrations in the cell relative to the extracellular fluid.

NA favors divalent metals ligand association with the NA-enzyme with the complex conformational changes mediated by turnover of H-bonds. The protein-hydration shell turnover of broken but reconstituted through equilibrium with water cluster depletion of participating H-bonds involves about 1-7kcal/mol. This exceeds the ambiance thermic energy of about 0.8kcal/mol permitting to adapt signal to noise ratios for efficiency of the neuronal synaptic communication. The frequency transduction of a pulsatile kinetic event allows that a neuronal frequency of about 20-40MHz could be associated into an output nerve impulse.

In the red cell-Hb during deoxygenation could discharge Mg^{2+} at the synaptic cleft. From a network perspective a near simultaneously presynaptic action potential opens the voltage-sensitive calcium channels in the presynaptic membrane.

At the initiation of an action potential, a channel for Na^+ open [200, 201] and the ion moves producing a change in polarity across the axon membrane. Repolarization results from the opening of the channel for K^+ which moves out of the axon. The impulse travels unidirectional down to the axon terminal.

Voltage gated Na^+-channels [199] allow action potential lasting 1 millisecond whereas calcium-based action potentials could last for about 100 milliseconds.

In cardiac muscle cells, an initial fast sodium spike is a *primer* that onsets rapid calcium spike for muscle contraction.

Membrane depolarization (ΔV) activates specific voltage-operated channel (VOC) isoform voltage operated channels (VOCs) in the plasma membrane to allow a rapid influx of external Ca^{2+} to provide a Ca^{2+} trigger that then activates the ryanodine receptor 2 (RYR2) to release Ca^{2+} stored in the sarcoplasmic reticulum (SR) through a process of Ca^{2+}-induced Ca^{2+} release (CICR). This mechanism is found in cardiac muscle and neurons.

This influx of calcium into the cytoplasm leads to neurotransmitter-vesicles releasing NA to the outside of the presynaptic cleft membrane. Impulse transmission and/or NA binding to its receptor induce the depolarization of the plasma membrane. Neuronal transmission is terminated by the neurotransmitter reuptake linked to Na^+ entry [200, 201]. The β-receptor is present in the hormone-binding subunits of the enzyme AC. When NA binds to its receptor, this one becomes activated to bind G-protein, which is activated by GTP. As has been reported the neurotransmitter activated enzyme, increases Vmax. A network perspective allows integrating Mg^{2+} activating the responsiveness to NA of the postsynaptically located AC. The activated enzyme generates endogenous cAMP, which has the capability to induce changes in the function of ion channels within the membrane.

Postsynaptically, the activated protein kinase-A (PKA) could trigger the ion Na^+ translocation across the voltage-gated Na^+-channel, starting an action potential an completed under the ions Na^+/K^+ cotransport coupled to the Na^+/K^+-ATPase-pump, restoring the initial transmembrane potential. The ATPase itself shows an ATP hydrolysis-dependent turnaround over 3 angles of 120° each coupled to phosphorylation-dephosphorylation of a critical Asp residue in the ATPase itself [82].

ATPase activity of intact cell membranes could manifest ADP-phosphorylation by reversal of K^+ vs Na^+ ion loading or uptake [123, 124].

Figure 29 illustrates the rotational properties of Na^+/K^+ ATPase linked by these ions exchange to the electrogenic properties of neuronal membranes.

Modulation of slow inactivation by neurotransmitters acting through G-protein-coupled receptors, PKA, and protein kinase-C (PKC) is a flexible mechanism of cellular plasticity controlling the firing behavior of central neurons.

Kosmotropes ions organize water into its hydration shell and have the tendency to subtract water from chaotrope ions like $[K(H_2O)_6]^+$. To the first group belong $[Na(H2O)_6]^+$ with hexagonal geometry in the first water layer and $[Mg(H_2O)_6](12H_2O)^{2+}$ with octahedral geometry in the first and second layers.

The *nascent* Mg^{2+} greater reactivity allows the ion to capture part of Hydration shell of $[Na(H_2O)_6]^+$. The less hydrated or nascent $[Na(H_2O)_3]^+$ could return to its normal hydration shell, by capturing water from $[K(H_2O)_6]^+$.

An unsaturated hydric shell of the Mg^{2+} ion is capable to subtract water from hexa-hydrated sodium ion, $[Na(H_2O)_6]^+$ to form $[Na(H_2O)_3]^+$, even if the more fully hydrated shell of Mg^{2+}, may loss water to a less hydrated specie of Na^+. The latter, chaotropic strength could capture H_2O from the kosmotropic $[K(H_2O)_6]^+$ to generate tetra-hydrated $[K(H_2O)_4]^+$ and tri-hydrated complexes $[K(H_2O)_3]^+$.

The figure 29 illustrates that the increase in the manifest charge of Mg^{2+} leads to the capture of H_2O from the shell surrounding Na^+ at the outside of the neurons and by increasing its manifest charge to move for its inclusion at the pores/channel of the membranes. This uptake of Na^+ opens the voltage-gated channel at the beginning of the action potential.

The nerve impulse allows Na^+ from the cerebrospinal fluid (CSF) to enter in the cytoplasm side of the membrane to manifest a positive charge. Charge rearrangement may allow that the Mg^{2+} interacting with the membrane could be released as a less hydrated ion. Breaking a coordinative Mg^{2+}-membrane complex is endergonic and its forming is exergonic. This effect could couple charge exchanges with energy transduction as electrons displace within the electrogenic membrane.

If changes in hydration shells are inferred for ion translocation at the Na^+-pump [123] and at the ATPase, these two processes may cross-couple for partial loss of the hydric shells of Na^+ and K^+ to adapt their sizes [234] to corresponding gate.

The Mg^{2+} discharged by Hb deoxygenation within the red cell could be characterized by its manifest much higher effective charge than the fully hydrated ion because during oxygenation Mg^{2+} to form a chelate with the R-group losses most of its hydration shell. Hence, when the oxyHb-Mg^{2+} and other Mg^{2+}-chelating metabolites complexes break or dissociate release Mg^{2+} surrounded only by an incompletely saturated hydric shell. The increment in the effective charge of Mg^{2+} allows the ion to seek water, altering endothelial cells tight junctions. The Mg^{2+}-discharge at the synaptic cleft resulting in simultaneous activation of the NA-receptor on adrenergic neurons and inhibition of the Glu-receptor of the glutaminergic neurons.

The sizing of the hydration shell of Na^+ is widely accepted as required for the ion to penetrate into its channel at the Na^+-pump in order to reach the interior side of the membrane. The tendency to complete ion hydration shells became synchronized for nerve impulse. In the interior surface of the membrane $[Na(H_2O)_3]^+$ should react with $[K(H_2O)_6]^+$ and the generated $[K(H_2O)_2]^+$ may be able to transit its channel [234]. The Mg^{2+}-driven hydric coupling for ionic translocation decreases the magnitude of ΔH.

The cAMP increment could initiate a voltage-gated nerve impulse because the tendency of the more reactive Mg^{2+} to displace the outside Na^{+} into pore channel.

The Na^{+}/K^{+}-pump by breaking ATP^{4-} into ADP^{3-} and phosphate (Pi^{2-}) increases the concentration of ionic Mg^{2+}. The latter, by activating AC links the nerve impulse with the cAMP activation of protein kinases and the cAMP response element binding proteins. Hence, the dominant role of ATP-cycle to support the energy consuming processes of brain allows metabolic feedbacks which operate as a unification backbone of the processes leading to long-term memory consolidation.

Vasoconstriction and vasodilation of adrenergic and glutaminergic neurons could be related to astrocyte-arteriole network interaction with Mg^{2+} and Ca^{2+} in support of neuronal function.

The dissipative states of water-protein structures [235] involve the break and reform of H-bonds, which occur within femtoseconds. If this were only a thermic event the molecule could have a non-local conformation change and a non-spatially directed vibrational state.

The ventral root potential (VRP) responses (A- and C-fiber) and changes in receptor sensitivity is involved to the phenomenon of enhanced central excitability and sensitization, which occur in clinical conditions of inflammatory hyperalgesia and neuropathic pain and significantly reduced by antagonists of NK1, NK2, and NMDA/NMDAr (N-methyl-D-aspartate) an ionotropic receptor of the neurotransmitter Glu [236].

The Role of Na^{+}/K^{+}-ATPase on Action Potential

The central role of the Na^{+}/K^{+} ATPase is reflected in the energy available to the cells for supporting its membrane potential which became equivalent to about 25% total energy consumption of human at resting conditions.

An enzyme with His-side groups shifting in a ΔpKa mode or similar for other residue R-groups in response of a protein changing ligands, could allow to model the turnover for hydrated states of ions and protein related to the dynamics ATPabse driving the Na^{+}/K^{+}-pump turnover [82, 123, 234]. ATP's competition for an Mg^{2+} chelated within the ATPase protein could be assumed to change the dipolar state of the ATPase. The ATPase activity may results in a coordinative phosphorylation, which may change electrostatic attractions into repulsions within the subunits of the K^+ and Na^+ carriers [124]. The ATPase-dependent changes in the state of attraction between subunits of the carriers result in the appearance of dynamic structural modifications, which allow synchrony between K^+ uptake and Na^+ extrusion by the cell. This also requires synchronization between availability of glucose and Mg^{2+} [237]. If the glucose generated ATP^{4-} exceeds the availability of Mg^{2+} the effect could destabilize the Mg^{2+}-dependent binding of receptors to the membrane.

The figure 29 shows ATPase activity coupled to active transport of cation. Ion translocation occurs by conformational changes coupled to ATP hydrolysis. The P-type ATPases function according to an alternating the access pathways to both sides of the membrane [238] before opening to either side in an alternating manner [84] two conformational states. Crystal structures of sarco/endoplasmic reticulum Ca^{2+}-ATPase-1 (SERCA1) a in several key conformational states reveal that this enzyme can bind a nucleotide throughout the entire reaction cycle, either in a catalytic (E1) or a modulatory (E2) mode [239, 240, 241].

The E1 couples a high affinity for ATP and the cation to be expelled from the cytoplasm. ATP phosphorylation of an Asp residue in the ion-bound of E1 state forms a high-energy E1P.ADP state with a subsequent conformational change to the functionally lower-energy E2P state.

This one is followed by E2 activating the ATPase for dephosphorylation, changing the affinities such as a decrease for the binding cation and an increase for the counter ions [242]. Binding of counter-transported ions to the cation-binding sites (or other events that stimulate closure of the exit pathway) induces re-occlusion and dephosphorylation of the transporter (E2P to E2). The counterions are then released during transition to the E1 state by a conformational change associated with the opening of an extra-cytoplasmic exit path.

This extensive rotation exerts a pull on the three transmembrane helices linking with the A-domain: A-M_1, A-M_2 and A-M_3 (Toyoshima et al., 2004), changing in the tilt and position to become separated from the remaining transmembrane helices: M_5... to M_{10} [243].

Repolarization results from the opening of the channel for K^+ to move out of the axon. The impulse travels unidirectional down to the axon terminal.

Chapter 8

The Hypothalamic-Pituitary-Adrenal Axis Control on the Psychosomatic Metabolic Network

The characterization of the noradrenaline-dependent overstimulation of the hypothalamic-pituitary adrenal axis could turn-on the fight-or-flight response.

The increment of adrenaline and cortisol shift body metabolism in the direction of depleting metabolic reserves like fats and releasing amino acids from proteins to support gluconeogenesis. Adrenaline could not cross the blood-brain barrier for a negative feedback, returning to the homeostatic metabolic and functional states of brain. Adaptive evolution may explain the lack of adrenaline feedback as a restriction to signaling, which allows the dominance of brain over its metabolic supporting tissue network. A molecular perspective could therefore explain the advantage of assigning to the brain unchallenging control, for maximizing the organismal efforts required for survival.

This brain pattern of control over metabolic supporting functions corresponds to a similar equivalent of the structuring of a psychosomatic unconscious, unable of a feedback reporting on the conscious level.

The Activatory Effect of Mg^{2+} on the Adrenaline Stimulated AC of the Tissue Network

Mg^{2+} concentration exceeding MgATP [164] substrate concentration allows the activation of adrenaline responsiveness of particulate preparation of AC from fat cells [244]. Similarly effect is that the free Mg^{2+} excess allows conditioning the response to IRTK.

Table 8. Equilibrium concentrations by increasing Mg^{2+}, at constant 1mM total ATP concentration. The table illustrates the procedure to obtain ionic concentrations by calculating the equilibrium of the ions when $MgCl_2$ changes from 0.1 to 1 mM at a constant total ATP concentration (ATP_T) [164, 165]. Calculated ranges to obtain the separated effect of Mg^{2+} and ATP^{4-} are not informed, but were used to plot the reported experimental conditions

Concentration at equilibrium (mM)					
$MgCl_2$	ATP_T	Mg^{2+}	MgATP	ATP^{4-}	$[ATP^{4-}]/[Mg^{2+}]$
0.1	1.0	0.008	0.084	0.91605	109
0.25	1.0	0.021	0.186	0.81360	38
0.5	1.0	0.042	0.314	0.68577	16
1.0	1.0	0.084	0.478	0.52181	6

The effect of lipolytic hormones was to shift the activatory Mg^{2+}-dependent concentration range show by the saturation curves of fat cell AC responsiveness to the added hormones.

The apparent kinetic constants for activatory stimulus at 50% concentration for total $[Mg_T]$ $S_{0.5(MgT)}$, shifted from a basal value of 16.5mM to 6.1mM in the presence of 50μM of l-adrenaline, and to 4.5mM in 40μg/ml of adrenocorticotropic hormone (ACTH), and to 6.8mM in 35μg/ml of Glucagon. Caffeine was used to inhibit the phosphodiesterase that breaks cyclic adenosine monophosphate (cAMP) into AMP.

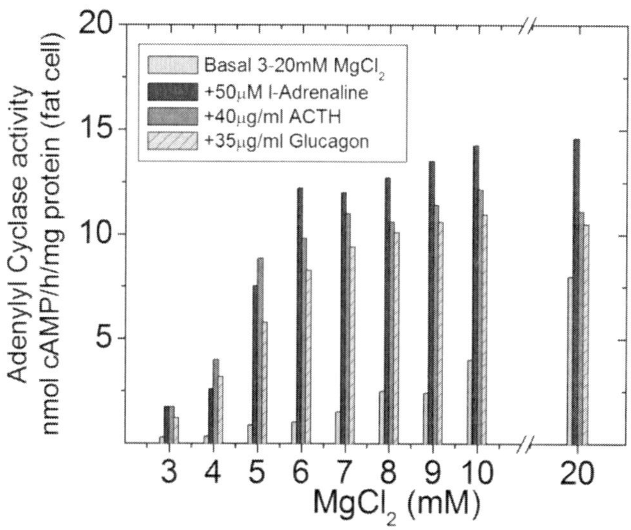

Figure 30. The Mg-dependent activation of fat cell AC-responsiveness to adrenocorticotropic hormone (ACTH), l-adrenaline and glucagon. 4.2μg of protein from a particulate preparation obtained from fat cells were incubated 10 minutes at 37°C according to methods: 50μl of reaction mixture with the inclusion of lmM-dithiothreitol, 5mM sodium-ATP (Sigma, horse extract with 2% nucleotide contaminants, GTP, etc.) at the indicated concentration of $MgCl_2$. Assay of cAMP, values obtained were plotted as bars. Basal values precede the ones for added hormone concentration. Values for enzyme activity: assays-(0-time blanks) reported in the vertical-axis.

The calculations for the species in equilibrium allowed to obtain saturation plots as a function of the concentration of the free ion Mg^{2+}, $S_{0.5[Mg^{2+}]}$ shifted from a basal value of 11.5mM in the presences of the lipolytic hormones to 1.2mM for adrenaline, 0.3mM for ACTH and 1.8mM for glucagon respectively.

The calculation allowed to separate the saturation curves of the inhibitor specie free ionic [ATP^{4-}]. Calculation for apparent 50% inhibition show that from $Ki_{[ATP4-]}$ basal 0.03mM decreased to 0.25mM in the presence of adrenaline, 0.62mM in ACTH, and 0.16mM in glucagon, respectively.

The figure 30 shows that activation by adrenocorticotropic hormone, adrenaline, and glucagon has a *common* dependence on free Mg^{2+}, in excess of that required to form MgATP, the substrate of AC [6].

The substrate MgATP concentration does not change by much and because it is comparatively large with regard to the other equilibrium species, it could be omitted from calculating overall equilibrium and allow to detect effects on the regulatory site.

The added hormones increase AC affinity for Mg^{2+} stimulation and decrease inhibition by ATP^{4-}, resulting in an additive magnification by the modulatory effects of small changes in the ionic equilibrium of the relative concentration of Mg^{2+} versus ATP^{4-}. The regulatory sites (RS) on the enzyme could be characterized for a positive modulation, by forming the enzyme RS-Mg^{2+} complex and for a negative modulation by forming the RS-ATP^{4-} complex.

The $MgCl_2$ increment at constant sodium-ATP concentration results in ionic equilibrium driving opposite changes. Simultaneously, increasing the saturation of the activator site and decreasing saturation of the inhibitor site. The separated equilibrium constants by incrementing [$MgCl_2$] show a relationship between both regulatory sites corresponding to K_1/K_2.

$K_1=[\uparrow RS\text{-}Mg^{2+}]/[\downarrow RS][\downarrow Mg^{2+}]$ and $K_2=[\downarrow RS\text{-}Mg\text{-}ATP]/[\uparrow RS\text{-}Mg^{2+}][\downarrow ATP^{4-}]$

This magnifies the hormonal effect:

$$K_1/K_2=[\uparrow RS\text{-}Mg^{2+}][\uparrow RS\text{-}Mg^{2+}][\downarrow ATP^{4-}]/[\downarrow RS]\,[Mg^{2+}]\,[\downarrow RS\text{-}Mg^{2+}\text{-}ATP^{4}].$$

Accordingly, the addition of $MgCl_2$ results in that the concentration of the activated state of the enzyme RS-Mg^{2+} increases to reach full saturation as a second order reaction rate, by the decreasing concentrations of both inactive states (RS and RS-Mg-ATP^{4-}) tending to null, $K_1/K_2 \propto [\uparrow RS\text{-}Mg^{2+}]^2$.

The values determined for cooperativity "n" for total were $MgCl_2$ basal $n_{(MgT)}=2.6$, which increases to 4.2 for adrenaline, 4.6 for ACTH and 3.5 for glucagon. The calculated separated concentration allowed to determine values for each of the equilibrium ion species.

Cooperativity for Mg^{2+} concentration shows for basal $n_{(Mg2+)}=1.2$ which increases slightly to 1.1 in the presence of adrenaline, 1.4 in the presence of ACTH and 1 in glucagon.

The corresponding separated Hill's constants for cooperativity to a decrease in ATP^{4-} concentration shows for basal $n_{(ATP4-)}=1.2$ which changes to $n_{(NA)}=1.1$ with adrenaline, 1.6 with ACTH and 1.0 with glucagon.

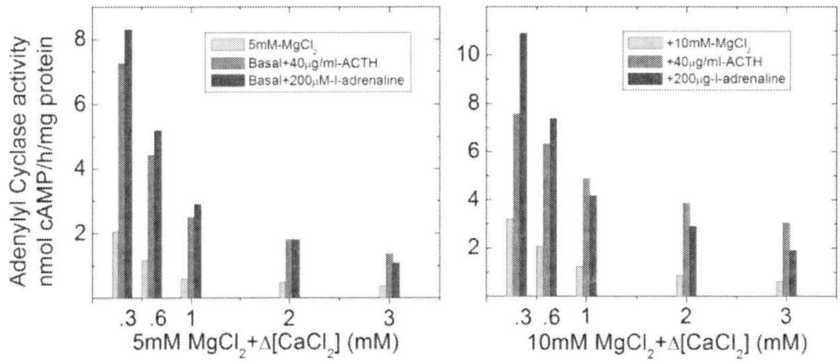

Figure 31. Inhibition by Ca^{2+} and CaATP of the Mg^{2+}-dependent hormonal responsiveness of AC. 3.7 μg of same particulate preparation from fat cells were incubated as in methods. In the presence of 5mM-sodium-ATP+5mM (or 10 mM) $MgCl_2$ as indicated in the horizontal-axis from 0.3 to 3mM $CaCl_2$. Hormone additions are indicated in the figures.

Effect of $CaCl_2$ on $MgCl_2$-Dependent Responsiveness to Hormones of Fat-Cell AC

The experimental conditions maintained allowed to isolate by theoretical calculation the effects of Ca^{2+} and CaATP to separate inhibitory and competitive role with regard to Mg^{2+}-dependent stimulation of AC by ACTH and l-adrenaline (AD) [6].

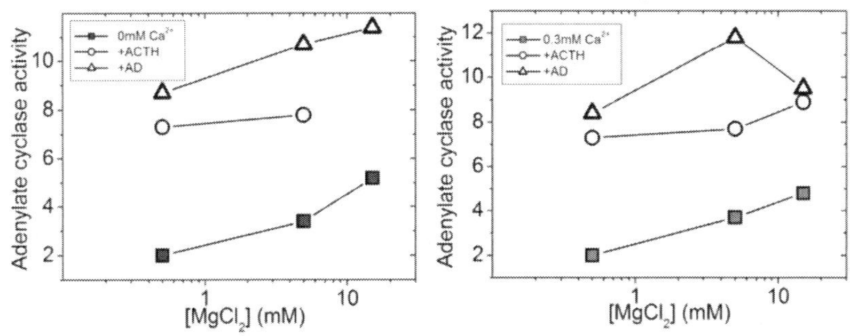

Figure 32. (Continued)

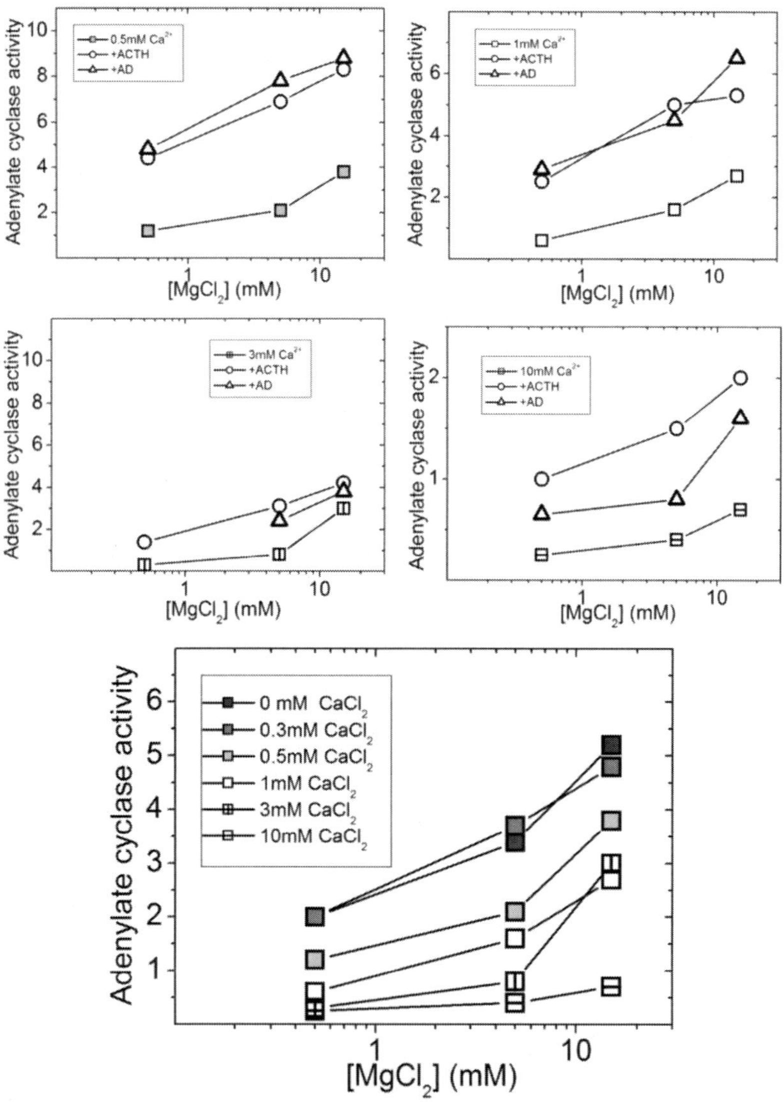

Figure 32. Effect of addition of $CaCl_2$ on 5mM-, 10mM-, 20mM-$MgCl_2$ dependent responsiveness to hormones of fat-cell adenylyl cyclase. Portions of an adenylyl cyclase particulate preparation containing 3.7μg of protein were incubated at 37°C for 10min in 50μl of a standard reaction mixture with the inclusion of 1mM-dithiothreirol, 5mM-sodium ATP. Hormones addition 200μg l-adrenaline (AD) and 40μg/ml ACTH and as shown in the subfigures (a) +5mM-$MgCl_2$ (b) +10mM-$MgCl_2$ (c) +20mM-$MgCl_2$ and increasing from 0-10 mM $CaCl_2$ ([Ca_T]) and solved for separated concentrations of ionic Ca^{2+} and CaATP as indicated in the plotting.

The metabolic effects of ionic equilibrium are similar to an amplification system, since large increases in the reaction rate of AC, occur by a change in chelating metabolite concentration like ATP^{4-}, $citrate^{3-}$, etc. This could lead to the capturing of free Mg^{2+} or releasing it, and also introduce large changes in the concentration of the inhibitory species ATP^{4-}, CaATP, Ca^{2+}, etc. The role of adrenaline in mobilizing fat stores could be replaced by other hormones over a different range of $[Mg^{2+}]$. If ATP^{4-} became exhausted increases the equilibrium level of free Mg^{2+}, allowing ACTH to became activatory at a much lower $[Mg^{2+}]$ concentration ranges, than the one for physiological activation by adrenaline and glucagon [6].

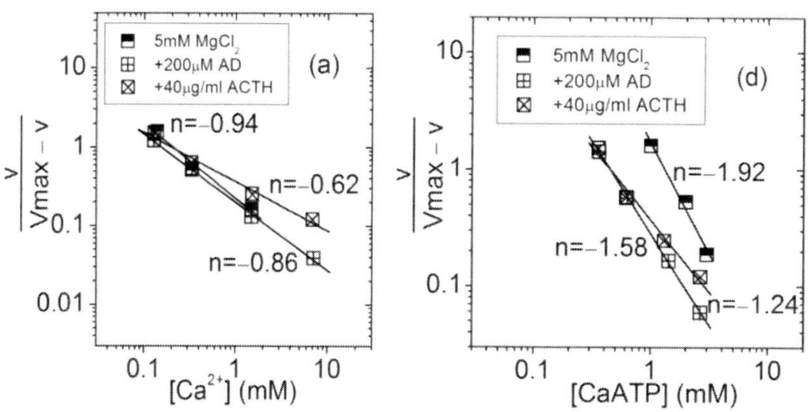

a) and d). Addition of 5mM $MgCl_2$

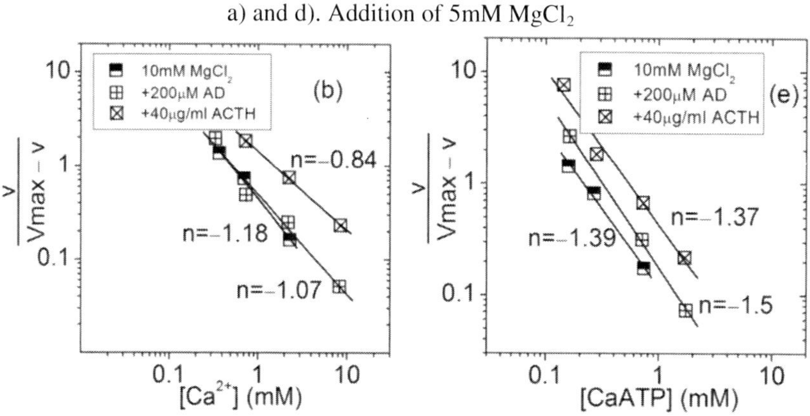

b) and e). Addition of 10mM $MgCl_2$

Figure 33. (Continued)

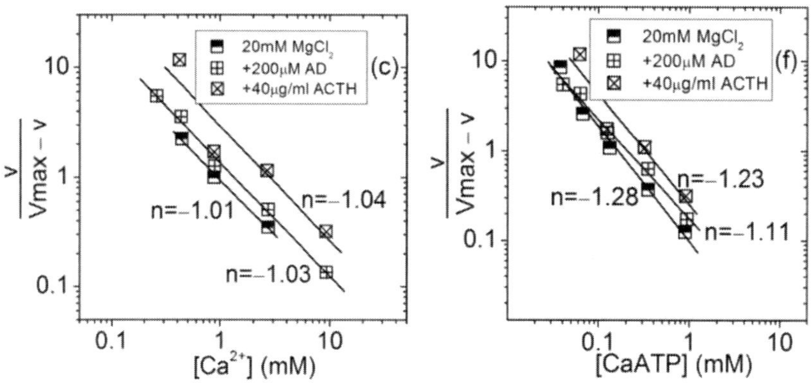

c) and f). Addition of 20mM $MgCl_2$

Figure 33. The separated Ca^{2+}- and CaATP-dependent inhibition of fat cell adenylyl cyclase. The effect of adrenaline and ACTH on releasing Ca^{2+} and CaATP-dependent inhibition of AC. The concentration (nM) of ionic calcium ([Ca^{2+}]) and calcium chelated to ATP ([CaATP]) were calculated utilizing the corresponding equilibrium equation and constants [164] and indicated directly on the logarithmic scales as it is shown on the figure.

Table 9. Effect of hormones on the AC apparent kinetic constants for Ca^{2+} and CaATP. The results obtained by increasing [$MgCl_2$] concentration were ploted according to Hill (1913) as a function of either ion Ca^{2+} or [CaATP] complex concentration. The cooperative values n(Ca^{2+}) and n(CaATP) were similar for 5mM, 10mM and 20mM added $MgCl_2$ [6, 8]

Constant determined for	$MgCl_2$ (mM)	Basal	l-Adrenaline (200μM)	ACTH (40μg/ml)
n(Ca^{2+})	5	0.90	0.85	0.60
	10	1.13	1.10	0.93
	20	1.05	1.06	0.98
Ki(Ca^{2+})	5	0.19mM	0.17mM	0.18mM
	10	0.49mM	0.59mM	1.55mM
	20	0.94mM	1.25mM	2.52mM
n(CaATP)	5	1.53	1.55	1.28
	10	1.34	1.30	1.23
	20	1.36	1.14	1.13
Ki(CaATP)	5	0.42mM	0.41mM	0.47mM
	10	0.19mM	0.15mM	0.45mM
	20	0.14mM	0.12mM	0.25mM

Figure 33 shows that [Ca^{2+}] was capable to inhibit AC even when Mg^{2+} was present at a much higher concentration, and adrenaline and ACTH only at the lower [Ca^{2+}] were capable to significantly activate AC [8].

The increase in $MgCl_2$ added resulted in increasing from 0.19mM to 0.49mM and 0.94mM $MgCl_2$ the value for 50% of inhibition Ki(Ca^{2+}). This effect may indicate competition between the ions Mg^{2+} activating and Ca^{2+} inhibiting for a single modulatory site [6, 8].

The table shows that at 5mM $MgCl_2$ show a similar Ki(Ca^{2+})=0.18mM for basal, l-adrenaline and ACTH-incubation conditions. The higher 10mM $MgCl_2$ by increasing Ki(Ca^{2+}) decreased inhibitory effect of the Ca^{2+} ions on all the mentioned incubation conditions with an stronger decreasing inhibtion in the presence of ACTH Ki(Ca^{2+})=1.55mM.

At 20mM $MgCl_2$ dependent decrease in the Ca^{2+} ion inhibitory effects was smaller for basal Ki(Ca^{2+})=0.94 than for l-adrenaline Ki(Ca^{2+})=1.25mM and ACTH Ki(Ca^{2+})=2.52mM.

A hormone-dependent decrease of Ki(Ca^{2+}) may correspond to the hormone-enzyme complex changing conformation at the binding site to manifest a greater affinity for Mg^{2+}, therefore decreasing affinity for Ca^{2+}. An decreasing the capability of Ca^{2+} for competitively displacing Mg^{2+}. The $MgCl_2$-dependent increase in Mg^{2+} ion could decrease the apparent Km of the enzyme.

At 5mM $MgCl_2$ the Ki(CaATP)=0.42 applied similarly to basal and hormonal incubation conditions at 10mM $MgCl_2$ the Ki CaATP of about 0.14, 0.12, 0.25mM decreased, showing that competition by Mg^{2+} was not capable to displace CaATP, wich also showed stronger inhibitory action than Ca^{2+} ions. Mg^{2+}-dependent decrease by the hormones of Ki(CaATP), inhibiting the enzyme could be related [8] to an conformation changes, affecting Vmax [6, 8].

The citrate overflowing from the mitochondria into cytosol is a metabolic condition which activates lipogenesis at high Atkinson's energy charge. The effect of citrate could be opposed by adrenaline, by the activatory effect based in ionic equilibrium effect of the Mg^{2+}-dependent responsiveness of AC. A Hill treatment of data shows that addition of 50μM l-adrenaline decreased Mg^{2+} saturation requirement for 50% velocity (v) $S_{0.5(Mg2+)}$ from basal 11.5mM to 1.2mM and increased inhibitor saturation requirement $Ki_{(ATP4-)}$ from basal 0.03mM to 0.25mM respectively.

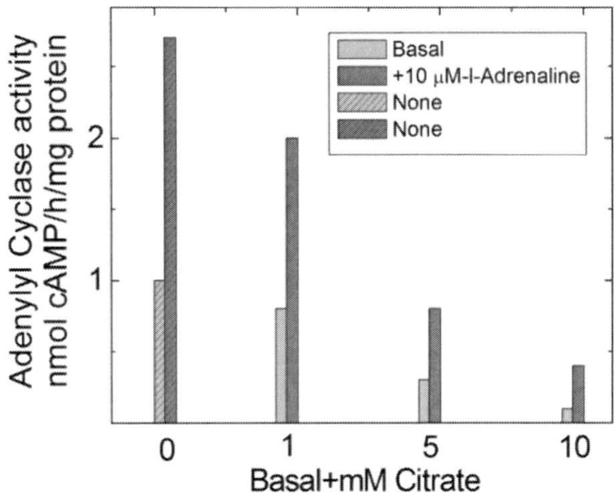

Figure 34. Citrate inhibition by competitive chelation of the Mg^{2+}-required for the basal and the adrenaline activation of adenylyl cyclase. 5.9 μg of adenylyl cyclase particulate preparation from fat-cell were incubated at 37°C for 20 min in 50μl of a standard reaction mixture which included 1 mM-glutathione (reduced), 0.1 mM-MgATP, 5 mM-$MgCl_2$, increasing concentrations of citrate. Zero time blanks were prepared and the cyclic AMP was assayed by the procedures described in Methods. Results are the means of four assays for each incubated condition.

The Hypothalamic-Pituitary-Adrenal (HTPA) Axis

Figure 35 illustrates that the G-protein coupled receptor (GPCR) is constituted by a transmembrane heptahelical protein, which pass through the cell membrane seven times (7TM) [245].

A large number of receptors are constitutive by 7TM, which binds hormone molecules outside the cell to endogenously activate signal transduction pathways and, ultimately, cellular responses.

There are 7TM receptors which recognize single atoms such as calcium, large pituitary hormones, morphine, histamine, and the ones that distinguish thousands of different smells and taste. There are receptors called orphan, because lack identification of the corresponding activatory hormone or neurotransmitter. About a 20% of marketed drugs interact with a GPCR.

The hormones are ligands at the loops binding G-proteins at the helices 5 and 6 of 7TM. G-proteins are heterotrimers complexes of three proteins: α, β, and γ subunits.

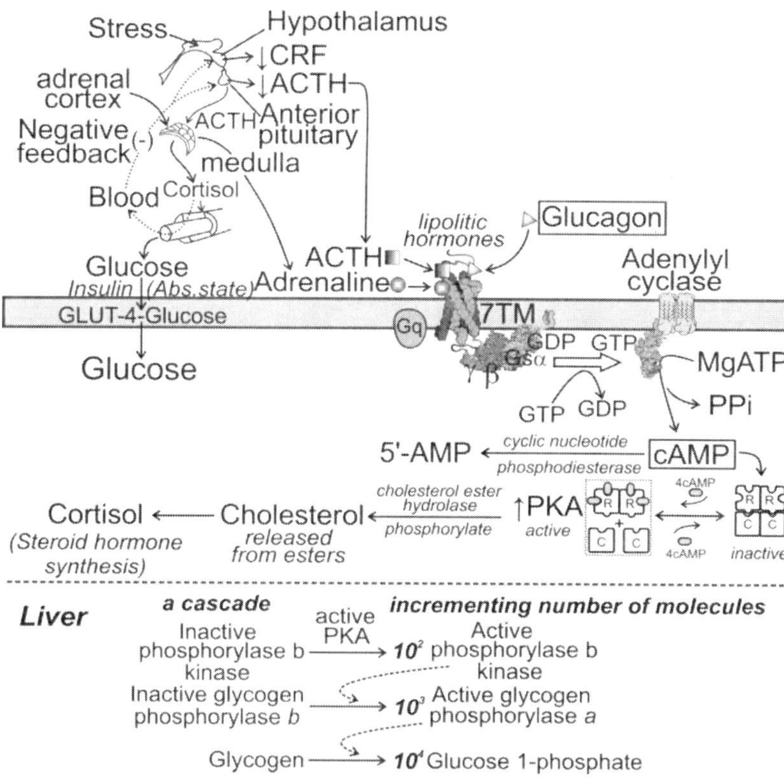

Figure 35. The HTPA axis and the responsiveness to glucagon, adrenocorticotropic hormone (ACTH), adrenaline of tissue network. Hypothalamus releases corticotropin factor (CRF) to stimulate the anterior pituitary to release ACTH. This one binds to the melanocortin-2-receptor on adrenocortical cells activating AC.

Protein kinase-A (PKA) is a tetramer of a regulatory RR and catalytic CC dimers. Cyclic adenosine monophosphate (cAMP) binds to the RR dimer of an inactive PKA to dissociate and activate each of the two C monomers. Thus, initiates phosphorylation cascades, illustrated for the in the liver pathway, which turns inactive glycogen phosphorylate-"b" into the active form phosphorylate-"a". The catalytic subunit of PKA can also move into the nucleus, where it phosphorylates a protein that alters the expression of specific genes. The isolation from tissues of adrenergic receptors yielded only a few copies, requiring a 10^5-fold purification [246]. A cascade like the illustrated of phosphorylase-*b* conversion into -*a* allows to magnify the hormonal response, because at each subsequent step, the implicated number of activated molecules became highly increased. In adrenocortical cells melanocortin 2-receptor couples to activate AC increasing cAMP an activate PKA. Figure 35 illustrates that AC generates cAMP activating PKA, at the adrenal cortex. Phosphorylation of the enzyme cholesterol ester hydrolase activates the release from cholesterol esters of the free cholesterol required for the synthesis of cholesterol-derived hormones like cortisol. Cortisol as glucuronide or associated to a protein could bind to a G-protein on the plasma membrane, but can transverse the brain-blood barrier (BBB) for feedback effects on the HTPA axis, inhibiting the release of corticotropin factor (CRF) and adrenocorticotropic hormone (ACTH).

Glucose Metabolism

The insulin-regulated transporter GLUT4 is found in adipose tissue and striated muscle. GLUT1 is present in erythrocytes and endothelial cells of barrier tissues like the Blood brain barrier (BBB), is also expressed in liver cells which require bi-directionality to uptake glucose and by having glucose 6-phosphatase to release it from glucose-6-phosphate (G6P). GLUT3 is expressed mostly in neurons. Brain and liver use transporters which do not require insulin for efficient uptake of glucose. Cells, particularly in skeletal muscle and the liver, store excess glucose as the polysaccharide glycogen, and quickly metabolize it again when necessary to support brain function. The pathological consequences of diabetes and obesity involve gene expression changes in glucose metabolic pathways. In one notable example, PCK1 overexpression in mice leads to obesity. Tumors often show decreased oxidative phosphorylation, even in the presence of sufficient oxygen, due to enhanced transcription of glycolytic genes and/or reduced transcription of TCA cycle genes.

The Adrenaline Control of the Energy Stores: Hepatic Glycogen and Adipose Tissue Triglycerides

Corpus coeruleus neuronal discharge of noradrenaline at the hypothalamus on the HTPA axis results in the raising of adrenaline in blood, signaling for the fight-or-flight response.

Target tissues in the presence of free Mg^{2+} respond to the hormonal stimulus of AC by incrementing endogenous cyclic adenosine monophosphate (cAMP). This second messenger shifts metabolic activity to decrease hepatic glycogen and adipose tissue triglycerides to respond to fight-or-flight response increased energy needs at target tissues. Aroused functions like the increase the heart beating and the blood availability to muscle require negative feedbacks to return to resting conditions.

Dynamic ionic balance integrates changes in the qualitative composition and levels of the intermediate metabolites, capable to modulate the concentration of free divalent metals increasing responsiveness to adrenaline, adrenocorticotropic hormone (ACTH) and glucagon on the activation of AC in fat cells membranes.

The ionic balance could signal most purposely that could do the depletion of Energy Charge (EC), which could not distinguish between a non-chelating metabolites like the substrate MgATP without effective charge, turning and chelating metabolites like free ATP^{4-} capable to inhibit the membrane enzymes AC and IRTK [164].

If the ionic equilibrium between divalent metal (Mg^{2+} or Mn^{2+}) with ATP^{4-} favors the increase of free ATP^{4-}, CaATP or Ca^{2+}, the result turn-off of the hormonal responsiveness of AC.

Adrenaline inhibits the pancreatic insulin release, transferring to the ionic equilibrium increasing Mn^{2+} or Mg^{2+} control of the IRTK responsiveness to insulin. Under these conditions the metabolic tendency will favor the flattening of the circadian rhythms, because ATP^{4-} accumulation could not signal for the activation of biosynthetic pathways, capable to fully restore the metabolic stores.

Hormonal signaling integrates with a chelate metabolite decrease signaling in feedback that a release of the activatory free Mg^{2+} induces an equilibrium decrease of ATP^{4-}, an inhibitor of AC. This, sum-up as a double activatory increased that magnifies the lipolytic hormonal responsiveness of AC, over its basal activity by increasing the enzyme affinity for Mg^{2+} and decreasing its affinity for ATP^{4-}. The adrenaline signaling shift decreases the synthesis of energy stores. This jointly with cortisol-activation of protein degradation to support gluconeogenesis leads from a non-sustainable metabolic hyperfunction of the fight-or-flight response at the initiation of stress, into the exhausted metabolic state characteristic of chronic stress.

The Glucocorticoid Receptor

The zona fasciculata of the adrenal cortex in response to stress releases a steroid hormone cortisol (cortisone or hydrocortisone) with the primary function of increasing blood sugar through gluconeogenesis.

Corticotropin-releasing hormone (CRH) and β-endorphin directly and indirectly through the action the glucocorticoids inhibits the release of gonadotropin-releasing hormone (GnRH). Cortisol inhibits the production of thyroid stimulating hormones (TSH). Chronic activation of the HTPA axis suppresses growth hormone (GH) release and inhibits the action of insulin-like growth factor-1 (IGF-l) at target tissues.

Cortisol diffuses through the cell membrane into the cytoplasm and binds to the glucocorticoid receptor (GR) that resides in the cytosol. GR regulates genes transcription expressed in several forms with differentiable effects on different tissues of the body. The GR is complexed with a variety of proteins it's binding of the hormone releases the heat shock proteins resulting in the activated form of GR.

GR transactivation involves the receptor, translocation via active transport into the nucleus, and binds to DNA responsive elements for specific activation of gene transcription.

Transrepression involves that the absence of activated GR, allows other transcription factors such as NF-κB (nuclear factor kappa-light-chain-enhancer of activated B cells) or AP-1 (activator protein 1) to transactivate target genes.

The action of cortisol could develop into a pathophysiology of the sympathetic nervous system eg, leading to chronic stress with syndrome of uncontrolled proteolysis nurturing gluconeogenesis, hypertension and autoimmune disorders.

The stimulation of cortisol release reaches its high during the early morning waking hours, and declines as the afternoon progresses. It occurs with a pulsatile frequency of 7-15 times per day and occurs within 15 minutes of the surge of adrenocorticotropic hormone (ACTH), following a circadian rhythm that is sensitive to light, sleep, stress, and disease. The augmented cortisol, disrupts food-intake, sleeps and interferes with the biological clocks of circadian rhythms.

The subtypes of G-protein-coupled receptor OX1-R [247] allows a cortisol-dependent stimulus on the adenylyl cyclase signaling cascade. After passive diffusion through the cells membrane binds the type II glucocorticoid receptor. This complex translocates into a nucleus binding to the sequences of glucocorticoid response elements (GRE) in the promoter of DNA, altering gene transcription.

The lipid soluble cortisol has access to the HTPA axis, which is not allowed to the water soluble adrenaline. The latter by flattening circadian rhythms of glucose and insulin could lead subject the HTPA axis to sense the lack of glucose.

This allows stress to induce large increases in the circulating cortisol. It is an over-simplification when it is not assess the role of corpus coeruleus-noradrenaline (NA) control on the adrenal medulla chromaffin cells production of adrenaline, as the primary cause of stress.

Resolution and reconstitution studies show that the binding of ATPase and AC to the membrane are mediated by Mg^{2+} [10]. AC is inactivated by overexposure to NA. Accordingly AC replacement by a newly synthetized enzyme requires the inactive specie to be released from the membrane. Capture of the Mg^{2+} participating in the binding of proteins within the membrane may require and ATP^{4-} concentration well in excess over that of substrate MgATP.

The Lack of Adrenaline Access to HTPA Axis Prevents a Negative Feedback for Metabolic Homeostatic Control

Adrenaline could not cross the brain-blood barrier (BBB) for a feedback turn-off of the hyperstimulated HTPA axis. Thus, allowing persistence of the stress-stimulus, acting on the hypothalamus. This one leads to a constant increase in cortisol, and thus becoming the main stress-hormone. This increases the release of orexins A and B. Orexin-A is capable to stimulate adrenocortical cells to secrete cortisol.

The Responsiveness to Lipolytic Hormones of Adenylyl Cyclase

Growth hormone (GH) let to the release of free fatty acids from adipose tissue [244]. The increased lipolytic sensitivity conferred on the cells by preincubation with GH and dexamethasone (Dex) could be correlated with changes in endogenous cyclic adenosine monophosphate (cAMP) levels. Adrenaline potentiated this effect. Theophylline-dependent inhibition of phosphodiesterase increased cAMP level and GH and Dex potentiate the effect of theophylline. GH and Dex were shown to potentiate the glucagon-dependent effect on cAMP levels [8].

Adrenaline could not transverse the brain-blood barrier (BBB) acts as a first messenger on its receptors located outside of the cell membrane [248] of fat cells of liver, etc.

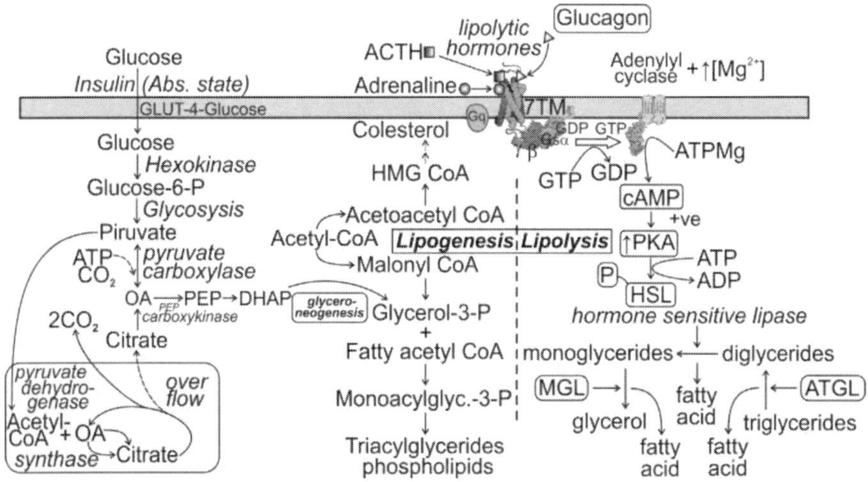

Figure 36. Glyceroneogenesis. The pathway is essentially an abbreviated version of gluconeogenesis. Pyruvate in the mitochondria is acted upon by pyruvate carboxylase to form oxaloacetate (OA), a citric acid cycle intermediate. Glucocorticoids suppress in adipose tissue the expression of the gene encoding phosphoenolpyruvate (PEP) carboxykinase (PEPCK), which in glyceroneogenesis converts OA to PEP.

The figure 36 illustrates lipolysis by action of the adipose triglyceride lipase (ATGL) selectively hydrolyzes the primary ester bond in the adipose tissue of mammals. Hormone sensitive-lipase (HSL) hydrolyses the second ester bond. The lipolytic hormones cascade by activating AC and protein kinase-A (PKA) allow the hormone sensitive lipase to generate glycerol and FFA.

Humans do not have an enzyme capable to generate glycerol-P from glycerol. Glycerol could not be used for the synthesis of triacylglycerol (fats). This prevents the recycling of fatty acids into triglyceride and the free fatty acids have to be released into the blood.

If PEP carboxykinase (PEPCK) a gluconeogenic enzyme is present in the fat tissue, the fat cell remains unable to synthetize glucose. Triglycerides (fat) generation is dependent of insulin stimulated glucose uptake from blood and glycolysis. In this pathway PEP became converted to dihydroxyacetone phosphate (DHAP), as the source glycerol 3-phosphate (Glycerol-3-P) required to support tri glyceride synthesis. Cortisol decreases synthesis of the Glycerol-3-P required for glyceroneogenesis and lipogenesis in adipose tissue.

The Metabolic Regulatory Role of Hormones

The pathways for lipogenesis and lipolysis could not function simultaneously. Insulin deficiency and high levels of counter regulatory hormones adrenaline, glucagon, and cortisol combine to increase the activity of hormone-sensitive lipase to increase the release of free fatty acids, and decrease the activity of acetyl-coenzyme A (CoA) carboxylase.

Cortisol and its synthetic replacement drugs prednisone and dexamethasone have a role in controlling well-being, immune cells, inflammation, blood pressure, and in the maintenance of connective tissues such as skin, bones and muscles.

The hormone is involved in fat, protein and carbohydrate metabolism, development, and the bone resorption [249] decrease in bone formation and suppresses the immune system.

Physiologically, the early morning increase in cortisol release increases muscle proteolysis, releasing into blood an amino acids equivalent to about 1% per day of the total muscle weight. Subsequently a release of testosterone reincorporates the amino acids in blood into muscle [249]. The heart responsiveness to cortisol allows the release of amino acids from its muscular structure, which could be a factor linking the time of maximal dead rate to a peaking on cortisol levels.

The released amino acids support liver gluconeogenesis and its output of glucose to maintain in the awaken individual the metabolic activity of brain.

In central nervous system structures the glucocorticoid receptor has interactions with noradrenaline and serotonin [250]. At the neural level has been implicated in short and long-term adaptation to stressors, including some mood dysregulations and dysfunctions like depression.

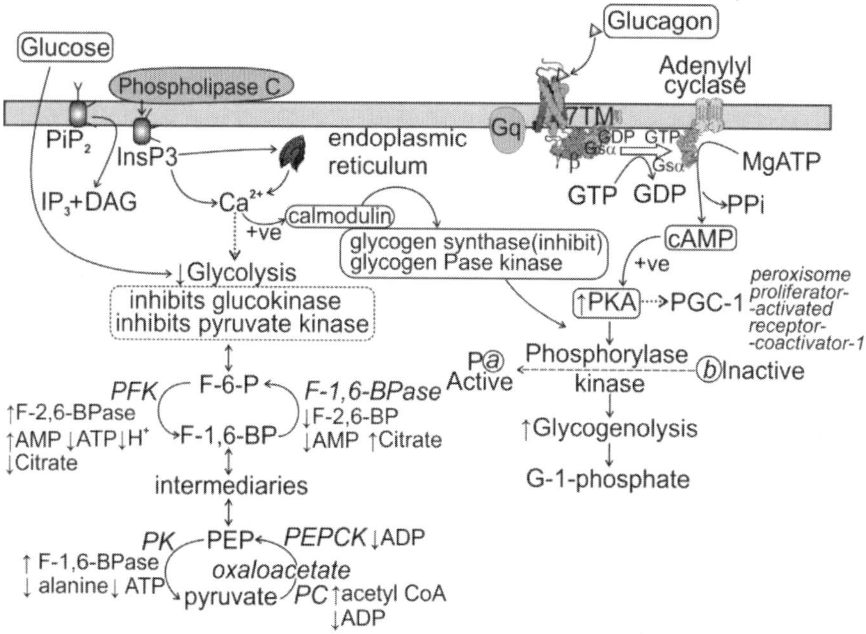

Figure 37. Enzymes in the metabolic network. Phosphoenolpyruvate (PEP), PEP carboxykinase (PEPCK), Glucagon binds to G-protein coupled receptor (GPCR) on main target liver, hepatocytes in which glucagon actives AC and increases cyclic adenosine monophosphate (cAMP) and protein kinase-A (PKA).

Protein kinase-A (PKA) activation involves phosphorylation of other enzymes, such as phosphorylase-b inactive to its active form phosphorylase-a. Dephosphorylation of the latter reverts to the form "b".

Liver has glucose-6-phosphatase (G-6-Pase), which allows the delivering of glucose by liver into blood. Peroxisome proliferator-activated receptor-coactivator-1 (PGC-1), decreases glucose entry into the glycolytic cascade -ve inhibits +ve activates.

Thirteen kinds of mammalian phospholipase-C (PLC) are classified into six isotypes (β, γ, δ, ε, ζ, η) and according to structure phosphorylate other molecules, leading to altered cellular activity. PLC cleave the phospholipid phosphatidylinositol 4,5-bisphosphate (PIP2) into diacylglycerol (DAG) and inositol 1,4,5-triphosphate (IP3).

IP3 functions by raising cytosolic [Ca^{2+}], which bind to its receptors like calcium channels in the smooth endoplasmic reticulum (ER). The Ca^{2+} concentration at the cytosol increases inducing a cascade of intracellular changes. Ca^{2+} and diaglycerol (DAG) are intracellular second messengers which activate protein kinase C.

The Insulin Receptor and the Hormonal Control Mechanisms of the Pancreatic Cell

Figure 38 illustrates the insulin membrane receptor (IRS) is a heterotetrameric 2α and 2β-subunits, linked by disulfide bonds with the insulin binding domain of α in the outer face of the membrane and two transmembrane β subunits, which contain the protein kinase. The α-chains have the insulin binding site, which triggers the receptor auto-phosphorylation on the three tyrosine residues, at the carboxyl terminus of the cytoplasmic β-chain domain located on its partner dimer.

The physiological effect of insulin is mediated by activation of protein kinases for phosphorylation and of phosphatases for the dephosphorylation. The activated receptor phosphorylates tyrosine residues of several insulin-receptor proteins IRS-1, IRS-2, IRS-3 and IRS-4, which facilitate the interaction of the receptor with intracellular substrates.

Epidermal growth factor (EGF) to activate its receptor EGFR activated by extracellular ligands linked to tyrosine kinases. Ion Mg^{2+}-dependent in excess of substrate MgATP is required for phosphorylation of the tyrosine residues at the cytoplasmic domain activates the receptor [14].

Insulin receptor tyrosine kinase (IRTK) is an auto-kinase that from ATP transfers a phosphoryl group to the hydroxyl group of three tyrosine (Tyr) residues, at the intracellular domain of the IRTK.

This activates the phosphorylating site capable to phosphorylate Tyr residues of other specific target proteins.

Insulin binds to its receptor on the surface of the plasma membrane of its target cells. The insulin-like growth factor receptor-1 (IGFIR) promotes cell growth and inhibits apoptosis.

Glycogen synthase kinase 3 (GSK3) phosphorylates serine and threonine residues and manifests a kinase activity for about forty different proteins in a variety of different pathways. It has been implicated in diabetes mellitus type 2, inflammation, cancer, bipolar disorder and Alzheimer's disease [251].

Key ligands for these pathways include the hormone insulin and insulin-like growth factors I and II (IGFI and IGF2). These ligands bind with different affinities to insulin receptor and IGFI receptor homo- and heteroduplexes.

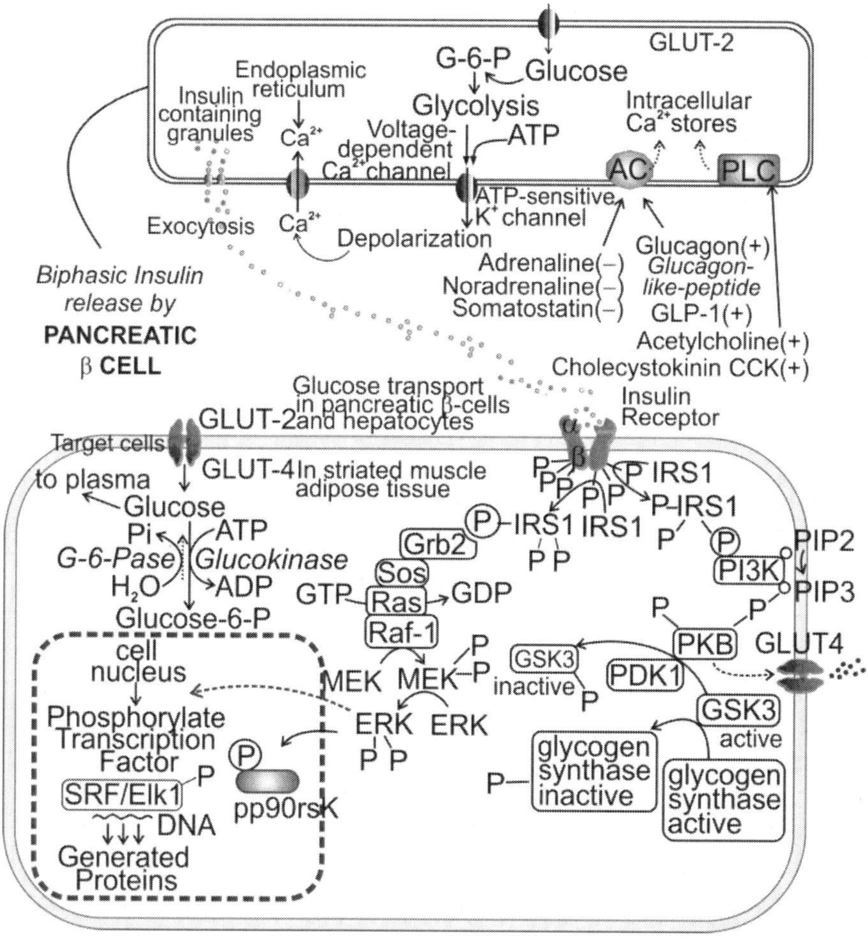

Figure 38. Pancreatic β cells and target cells in triated muscle adipose tissue. This tissue releases insulin in response to amino-acids and glucose. It is integrated with AC action, inhibited by adrenaline (-), noradrenaline (-), somatostatin (-), stimulated by glucagon (+), glucagon-like peptide-1 (GLP-1) (+) and the enzyme phospholipase-C (PLC), stimulated by acetylcholine (+), cholecystokinin (CCK) (+) allow the steering of insulin secretion to adjust the metabolism to a varying energy intake and expenditure. Intracellular inhibition of the 90 kDa ribosomal S6 kinase (pp90RSK) cascade reduced Low-Density Lipoprotein (LDL) receptor induction.

The major insulin-responsive transporters in striated muscle and adipose tissue are glucose transporter-4 (GLUT4) [252] and insulin receptor substrates (IRS), which favors activation of glucose uptake and protein synthesis.

The tyrosine protein kinase receptors (TRK) are present in all major nerve growth factor (NGF) targets, including sympathetic, trigeminal, and dorsal root ganglia [253]. There are the isoforms of TRK: A, B and C, which mediate neurotrophins effects.

The brain-derived neurotrophic factor (BDNF) and neurotrophin-4 (NT4) specifically activate the TRKB. Transcripts encoding the TRKC receptor are found throughout multiple structures of the central and peripheral nervous system. A TRKC-deficient mouse lacks most sympathetic neurons and all muscle afferents.

Signal transmission by phosphatidylinositol-3-kinase (PI3K) involves protein kinase-B (PKB), which becomes activated by phosphorylation dependent of phosphoinositide-dependent kinase-1 (PDK1). PKB (also known Akt) is a serine/threonine-specific protein kinase, which plays a key role in multiple cellular processes, such as glucose metabolism, apoptosis, cell proliferation, transcription and cell migration.

Cells could organize into multicellular assemblies through cell adhesion molecules. These proteins manifest weak interaction between themselves.

PKB activation will also lead to phosphorylation and inhibition of GSK3. This functions as a regulatory protein of glycogen synthase (GS).

Wnt signaling involves GSK3 sequestration inside endosomes, if GSK3 is inhibited by PKB phosphorylation prevents the performing of GTPase activation protein (GAP) function on Eukaryotic initiation factor 2 (eIF2). Its activity remains high and results in reduced translational initiation. In humans the EEF2 gene encodes for the protein Eukaryotic elongation factor 2 (eEF-2). This one, by insulin action, is rapidly dephosphorylated via activation of protein phosphatase 2A (PP2A). Taken together, reduced eEF2K-mediated phosphorylation and increased eEF-2 dephosphorylation lead to increased protein synthesis [254].

Cells have several sensory systems which function as nutrient-sensing pathways (AMP kinase, mTOR, PAS kinase, hexosamine biosynthesis and Sir2).

Figure 38 illustrates intermediate effects of insulin, activation of glycogen synthase (GS), pyruvate dehydrogenase (PDH), and inactivates phosphofructokinase II and hormone-sensitive lipase (HSL).

The metabolic network involves target tissues: muscle, fat, and liver [138, 159, 255]. In these promotes glucose and fatty acid storage stimulating glucose uptake, glycolysis, glycogen synthesis, gluconeogenesis, lipogenesis, protein synthesis and fatty acid oxidation. In these target tissues also inhibits lipid and carbohydrate degradation. Thus, stress-mediated redirection of nutrient uptake could lead to a pattern of metabolic deprived tissues.

MicroRNAs-mediated regulation of translation expressed in liver, adipose, pancreatic β-cells and skeletal muscle are upregulated and could act as feedback loops to regulate insulin signaling.

Extracellular Signal-Regulated Kinases (ERK) and Mitogen-Activated Protein Kinases (MAPK) Pathways

Figure 39 shows that signal transduction and the internalization of the receptor occur as described in figure 38 by activation of the phosphoinositide 3-kinase (PI3K). Signal transmission involves protein kinase-B (PKB/Akt), affecting CREB (cAMP Responsive Element Binding Protein), a transcriptional and coactivator, with a role central to cell growth and development.

The mitogen-activated protein kinases (MAPK) activity could be reversed by MAPK-phosphatase [256, 257].

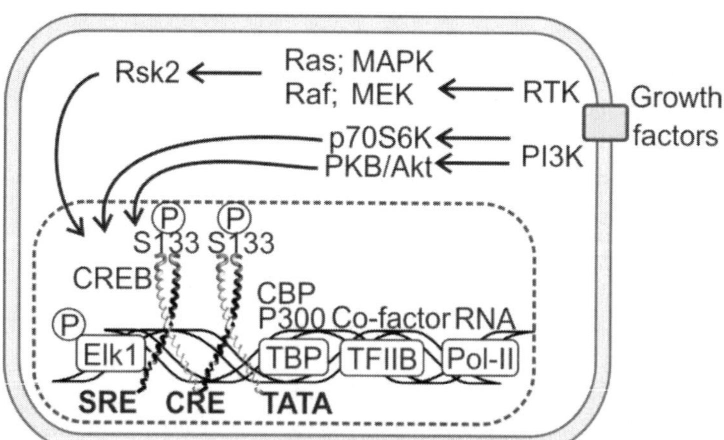

Figure 39. MAPK signaling system. Phosphorylation of Elk1 a component of a TCF (ternary complex factor) increases its transcriptional ability to form ternary complexes activates ribosomal s6 kinase (RSKs) binds serum response factor (SRF) to the serum response element (SRE) in the promoter region of many genes.

Figure 39 illustrates CBP/P300 (CREB binding protein/P300) acts by interaction with components of the general transcriptional machinery or by the acetylation of specific lysine residues in nucleosomes located near active promoters. Thus, creates access to the gene for the basal transcriptional machinery, TBP (TATA-binding protein). TFIIB (transcription factor-II-B) and RNA Pol-II (RNA Polymerase-II) stimulate expression of genes [258].

MAPK responds to signals from outside the cells for downstream activation of changes in target gene expression, promoting growth and differentiation effects. In liver is present the enzyme glucose-6-phosphatase (G6Pase) that releases glucose to blood.

MAPK signaling system has pathways that control different cellular processes like gene transcription metabolism, motility, cell proliferation, apoptosis, synaptic plasticity and long-term memory [259].

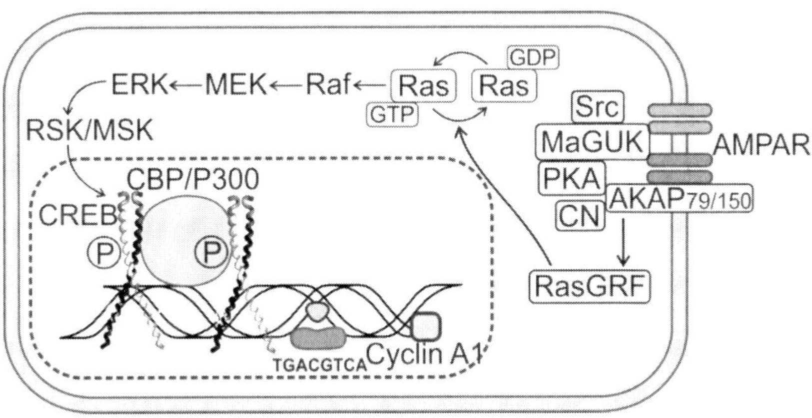

Figure 40. Extracellular signal-regulated kinases (ERK) pathway. It consists of a chain of proteins also denominated the Ras-Raf-MEK-ERK, Membrane-associated guanylate kinases (MAGUK), A-kinase anchor proteins 79/150 (AKAP79/150), α-amino-3-hydroxy-5-methyl-4-isoxazolepropionic acid receptor (AMPAR), calcineurin (CN).

Gene expression and protein synthesis, mediated by the transcription factor CREB (cAMP response element binding protein), play an important role in learning and memory in several species, including Drosophila, snails, and mice. Patients with the X-linked disorder Coffin-Lowry syndrome (CLS) have cognitive disabilities, distinctive features, and bony abnormalities as well as mutations in RSK2 (ribosomal S6 kinase-2), a protein kinase that activates CREB by phosphorylation at serine 133. In fibroblasts from a single patient with CLS, epidermal growth factor (EGF)-stimulated CREB phosphorylation was reduced. This suggests a correlation between human cognitive performance and cellular capacity to activate RSK2. It provides additional evidence that the CREB kinase, RSK2, and CREB phosphorylation may play important roles in human learning and memory, as they do in lower animals.

AKAP79/150 it is predominantly expressed in cerebral cortex, the encoded protein binds to the RII-beta regulatory subunit of PKA, for regulation of postsynaptic densities and events. Also binds for regulation PKC and phosphatase calcineurin and disrupts calcineurin-dependent dephosphorylation of nuclear factor of activated T-cells NFAT.

AMPAR is a non-NMDA-type ionotropic transmembrane receptor for glutamate, and at this the GluR2 (GluA2) tetramer forms glutamate receptor ion channel.

The receptor for glutamate GluA2 AMPARs is the most commonly found receptor in many parts of the brain. It has the ability to be activated by the artificial glutamate analog AMPA and participates in fast synaptic transmission at the central nervous system (CNS).

The MAPK/ERK pathway in nuclear transcription involves an activated ERK to move into the nucleus and phosphorylates the nuclear transcription factor Elk1 for transcription of about 100 insulin-regulated genes.

These proteins signal by adding phosphate groups to a neighboring protein, which act as an "on" or "off" switch.

Phosphorylated IRS-1 binds to growth factor receptor-bound protein 2 (Grb2), then to rat sarcoma (Ras), causing guanosine diphosphate (GDP) release and guanosine triphosphate (GTP) binding to Ras. Ras is 'switched on' by incoming signals and subsequently switches-on other proteins. Activated Ras binds and activates proto-oncogene c-RAF (Raf-1), a member of MAP kinase kinase kinase family. This phosphorylates the mitogen-activated ERK-activating kinase (MEK) on two serine residues, to activate it into phosphorylating ERK on Thr and Tyr residues.

Figure 41 illustrates the role of CREB as a nuclear transcription factor of genes, containing the cAMP responsive elements (CRE sites) involved in hippocampal long-term potentiation. It is a member of activating transcription factor (ATF) and cAMP response element modulator (CREM) belong to bZIP (basic/leucine zipper), which have been linked to human cognitive disorders [260].

Proteins belonging to this class are characterized binding to the palindromic sequence TGACGTCA.

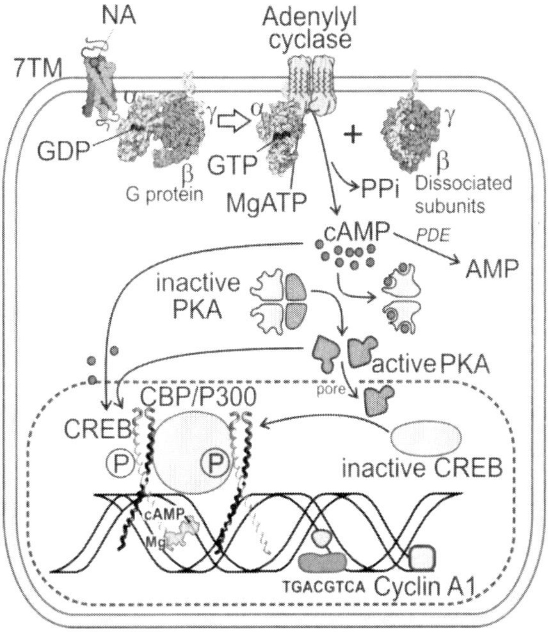

Figure 41. Noradrenergic neuron. Phosphodiesterase (PDE), CREB binding protein (CBP), CREB binding element, activated phosphorylated (aCREB), inactive CREB (iCREB). CREB protein recognizes the cAMP-responsive elements (CRE). The initial transcription site is coded either as a palindrome (TGACGTCA) or a half-site (CGTCA/TGACG) denominated as the TATA box motif at gene promoters.

This contains a leucine zipper for DNA binding (basic region) and for dimerization (leucine zipper region) of the proteins. CREB can form homodimers or heterodimers with other members of the activating ATF family, including ATF1 and CREM. However, heterodimerization of CREB with other members of the ATF family decreases its stability and CRE binding affinity.

CREB protein recognizes CRE, which occur either as an 8-bases palindrome TGACGTCA or a half-site CGTCA/TGACG at gene promoters.

Upon stimulation of cellular G-protein coupling receptor GPCR and growth factor receptors, adenylyl cyclase (AC) is activated by G-proteins: GN-alpha, GN-beta and GN-gamma increasing cAMP level. PKA formed by 4 subunits is located in the cytoplasm.

The binding of one cAMP per each of the subunits within the regulatory dimer protein, results in dissociation of two active catalytic subunits.

The hormone-bound activated receptor must be reset to prevent the continuous activation of G-proteins. Control is accomplished by hormone dissociation and by the bound GTP hydrolysis by Gα-GTPase activity.

The TATA box insertion restored cAMP responsiveness on a TATA-less promoter subsets of transcription-target genes are differently regulated according to core promoter configuration.

Changing levels of cAMP, Ca^{2+} and TGF-beta regulate CREB and its closely related proteins, which are implicated in a variety of biological responses such as neuronal excitation, long-term memory formation, neural cell proliferation and opiate tolerance.

CREB a central activator is generally constitutively expressed and can integrate cross-talk between signaling pathways through a variety of stimuli resulting in posttranslational modifications.

CREB through interaction with its nuclear partner CBP drives the transcription of a large number of genes. Phosphorylation by several different protein kinases has the potential for the convergence of multiple intracellular signaling cascades.

Activation of CREB occurs by the phosphorylation of Ser133 in the Kinase-inducible domain (KID). At this site phosphorylation of CREB can be caused by electrical activity. Several stimulus-dependent protein kinases PKA, PKC, CaMKs, CSNK, GSK3 and p70S6K can phosphorylate CREB. Protein phosphatase 1 (PP1) has been shown as the predominant phospho-CREB phosphatase.

CREB protein has been implicated in learning and memory, also functions in transcription requiring processes like growth-factor-dependent cell survival and glucose homeostasis [261].

Antidepressant CREB induction in the hippocampus could contribute to beneficial therapeutic results. According to the brain region activation of CREB could also show opposite roles.

Detrimental effects by CREB mediate stress, drug addiction, and depressive and anxiety-like behaviors, in the nucleus accumbens and several other regions [262]. The CREB protein has been shown overexpressed in leukemia. MicroRNAs act as negative regulators by the repression of ribosome decoding translation of RNA into a peptide sequence.

Promoters in a cell type–specific manner control the expression of genes: nuclear factor-κB, BLC2, cyclin A1/B1/D, JAK1-activated and signal transducer and activator of transcription-3, as many downstream protein kinases. These are involved in cell proliferation, differentiation, and survival signaling pathways: Akt/mammalian target of rapamycin and extracellular signal-regulated kinase. PLC catalyzes the cleavage of membrane lipid phosphatidylinositol 4,5-bisphosphate (PIP2) into the two second messengers DAG that remains in the membrane and inositol 1,4,5-triphosphate (IP3). The latter is soluble and diffuses away from the membrane and binds the endoplasmic reticulum (ER).

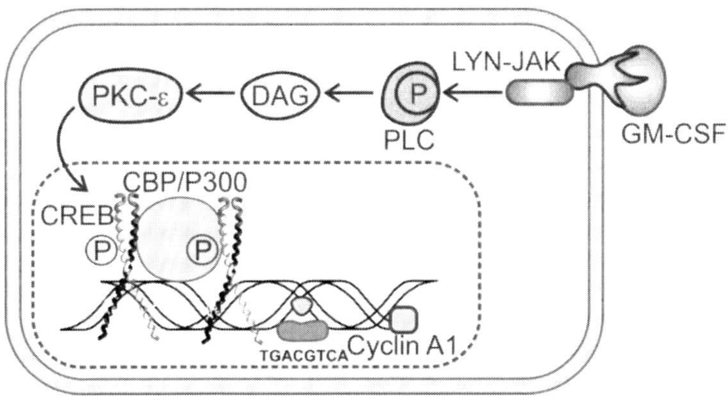

Figure 42. CREB target gene activation and PLC pathway, expression in haematopoietic cells. Granulocyte-macrophage colony-stimulating factor (GM-CSF), Lyn acts by inducing the Raf/MEK/ERK cascade, Janus kinase (JAK), phospholipase-C (PLC), 1,2-diacylglycerol (DAG), protein kinase C-ε (PKC- ε), Cyclic A1 in DNA repair, CREB Binding Protein (CREBBP or CBP) and p300 by phosphorylating Elk1 [263].

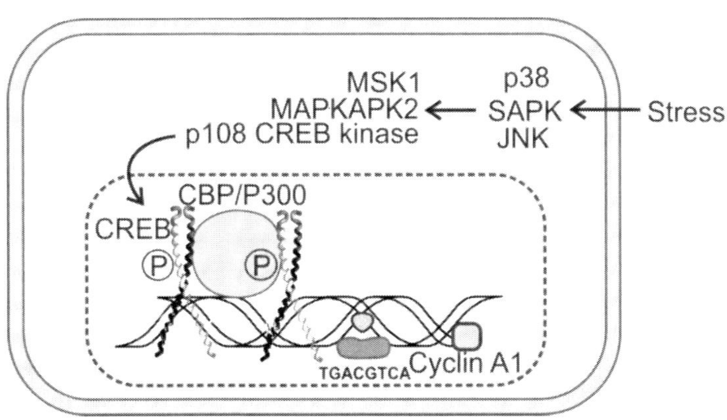

Figure 43. MAPK-JNK system. Their downstream effectors are activated by the MAPK components associated to extracellular-signal-regulated kinase (ERK) pathway, c-Jun N-terminal kinase (JNK) pathway and p38 pathway, stress-activated protein kinase (SAPK).

IP3-gated Ca^{2+} channel causes the rapid release of Ca^{2+} from the intracellular stores into the cytoplasm. A signaling Ca^{2+} binds calmodulin and PKC that phosphorylates serine and threonine residues of many target protein.

Calcium ion levels are kept low at about 0.1 μM by transport systems that extrude Ca^{2+} from the cytoplasm. Ca^{2+} coordinates strongly with 6 to 8 negative oxygen atoms. This includes the ones in the glutamate and aspartate side chains of proteins. The binding by the ion could result into cross-link of different segment of a protein, inducing substantial structural rearrangements.

SAPK/JNK is activated by many types of cellular stresses and extracellular signals, including embryonic development and immune responses. SAPK/JNK activation plays essential roles in organogenesis during mouse development by regulating cell survival, apoptosis regulation, and proliferation. Two SAPK/JNK activators, SEK1 and MKK7, are required for fetal liver formation and full activation of SAPK/JNK, which responds to various stimuli in an all-or-none manner.

Figure 44 illustrates those neurotransmitters, growth factors, hormone action on GPCR or by neurotrophin effects and receptor tyrosine kinase (RTK). CaMKIV, MAPKAP K2, and members of the pp90RSK (RSK), and MSK family could converge to phosphorylate CREB. The latter, in response to various signals and the figure shows how is diversely achieved by differentiable signaling pathways.

C/EBPs, first discovered in the induction of the acute-phase response in the liver, are also fundamental for long-term memory formation. As pointed out earlier, C/EBP and CREB family members participate in a fundamental, evolutionarily converged cascade of events that are required for the consolidation of new memories.

Figure 44 shows that the effect of neuronal activity on dendritic development is mediated by calcium influx via voltage-gated calcium channels (VGCCs) and NMDA receptors [265], as well as release from internal stores.

CaMKII functional heteromers the α to β isoforms, ratio regulates its association with the actin cytoskeleton and its microdistribution.

CaMKII strengthens synapses and restricts/stabilizes dendritic growth of cortical neurons in vitro.

CaMKIV, mainly localized in the nucleus, regulates dendritic morphology and responds to cytosolic calcium transients levels, entering into the nucleus to activate transcription events.

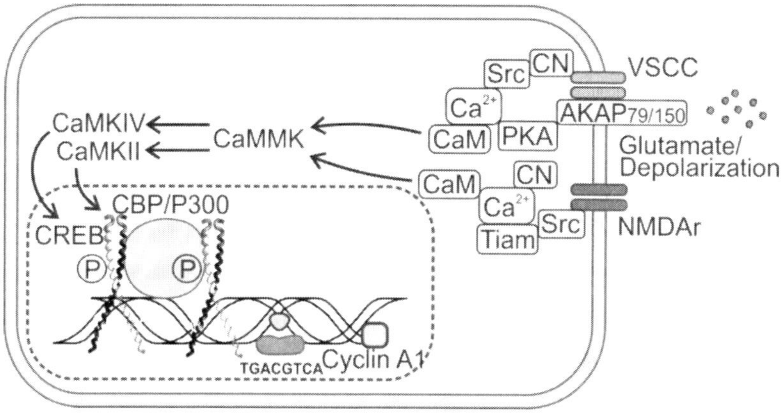

Figure 44. Ca^{2+}-dependent transcription. The voltage-sensitive Ca^{2+}-channel (VSCC), N-methyl-D-aspartate (NMDA) receptor, Ca^{2+}/calmodulin kinase II (CaMKII) isoforms, αCaMKII and βCaMKII, are present at high levels in dendrites; Ca^{2+}/calmodulin-dependent (CaMMK) [264].

Gluconeogenesis in the Brain Metabolic Network

Figure 45 illustrates that the catabolic hormones glucagon, adrenaline, noradrenaline and growth hormone act primarily, through cyclic AMP (cAMP) activation of protein kinase-A (PKA). Their action is countered by the anabolic hormone insulin. The latter, by promoting the increase in energy stores could deplete intermediate metabolites, at times of metabolic excesses.

The mitochondrial OA by the action of the mitochondrial PEP carboxykinase (PEPCK) converts into cytosolic PEP. The mitochondria OA by the action malate dehydrogenase became reduced to malate. The latter traverses the inner mitochondrial membrane. OA through transamination to aspartate enters into the cytosol.

The signals by glucagon or adrenaline lead to phosphorylation and inactivation of pyruvate kinase (PK) in liver, which increases the flux through gluconeogenesis. PK is also allosterically inhibited by ATP and alanine. These signal the adequacy of the level of substrates for gluconeogenesis. A reduction in energy levels became evidenced, by increasing concentrations of ADP. This, leads to inhibition of pyruvate carboxylase (PC) and PEPCK. The latter enzyme increases in starved individual in response to prolonged glucagon stimulation.

Expression of the key gluconeogenic genes for synthesis of G6P phosphatase (G6Pase) and PEPCK is dependent on plasma insulin concentrations. These do not change by much, but became expressed within the intestinal cells in between 24/48 hs after initiation of fasting. This allows the small intestine to contribute its endogenous glucose production. The promoter region of the G6Pase gene constitutes a binding site for transcription factors related to the homeobox (CDX) family.

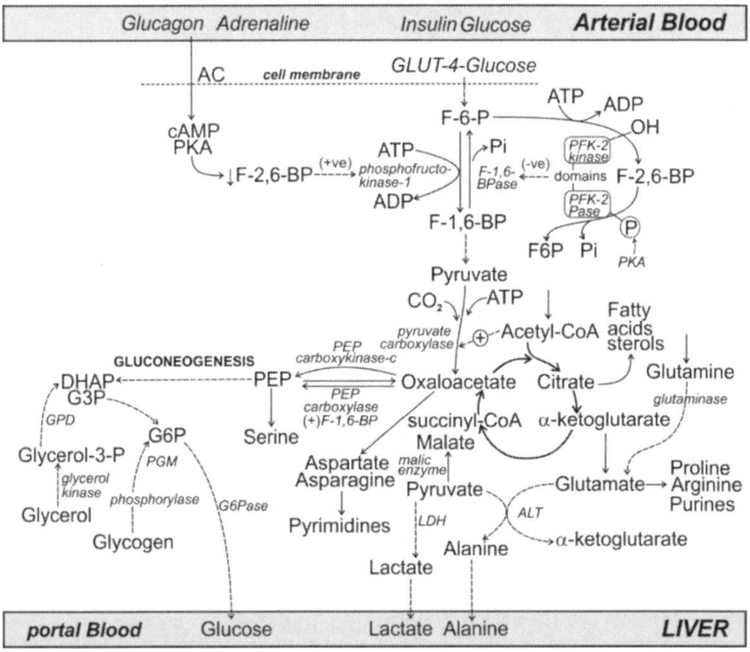

Figure 45. Anaplerotic reactions that form intermediates for pathways. The KREBS cycle function in anabolism (solid line). In liver/kidney, a deficiency in oxaloacetate let pyruvate carboxylase activated by (+)-modulator: acetyl CoA to catalyze in the presence of ATP to fix CO_2 to carboxylate pyruvate to OA. The increment in OA stimulates the cycle increasing incorporation of more acetyl CoA to form citrate. PEP carboxylase is activated by fructose 1,6-bisphosphate (F-1,6-BP). Glutamine by action of glutaminase and alanine transaminase (ALT) became α-ketoglutarate, which converts to OA and then to PEP to became a substrate for liver gluconeogenesis. Glycerol and glucose from glycogen of the intestine could contribute to the overall glucose homeostasis. Enzymes: Glycerol-3-phosphate dehydrogenase (GPDH), phosphoglycerate mutase (PGM), glucose-6-phosphate (G6P), glyceraldehyde-3-phosphate (G3P).

Controls fail when the body's hormonal secretion responsiveness became uncoupled from blood glucose levels. Lack of insulin secretion increases secretion of glucagon, growth hormone and catecholamines, activating lipolysis and hepatic gluconeogenesis. Dysfunction of hepatic insulin receptor tyrosine kinase (IRTK) causes the development of type 2 diabetes. However, the molecular mechanism regulating IRTK activity in the liver remains poorly understood [266].

Via the portal circulation intestinal enterocytes deliver, either from the diet or the arterial blood a supply of glutamine to liver. Glucose that enters the enterocyte by glycolysis became oxidized to pyruvate and transaminated to alanine. Lactic dehydrogenase (LDH) reduces pyruvate to lactate, which serve as major gluconeogenic substrate in the liver.

The acetyl-CoA allosteric activation (+ve) of PC occurs on a short time scale, whereas a long-term regulation can be exerted at the level of PEPCK.

Renal gluconeogenesis during periods of severe hypoglycemia provides glucose to liver. Glutamine is produced in high amounts by skeletal muscle, during periods of fasting to export the waste nitrogen resulting from amino acid catabolism.

In diabetes gluconeogenesis in the liver progresses without alteration by reduction or loss of the insulin signal, releasing sugar in spite of high blood glucose levels. Massive amounts of fatty acids are released to the circulation and the liver converts these to ketone bodies.

The high blood glucose levels lead to diuresis with loss of water, Na^+, K^+ and glucose, while the "ketones" lead to a pronounced fall in blood pH, in the absence of an effective treatment.

Negative Modulators of Glycolysis Act as Positive Controls on Gluconeogenesis

The regulation of gluconeogenesis will be in direct contraposition to the regulation of glycolysis. The shift is controlled by the activity of a bifunctional protein: phosphofructokinase-1/fructose 1,6-biphosphatase (PFK-1/F-1,6-BPase). In hepatocytes the level of the modulator (+ve) F-2,6-BP decreases in response to glucagon and catecholamines stimulation, through activation of cyclic adenosine monophosphate (cAMP)-dependent protein kinase-A (PKA).

PFK-2 kinase using ATP phosphorylates F-6-P to generate F-2,6-BP, which is an inhibitor (-ve) of the enzyme F-1,6-BPase. The substrate F-2,6-BP is the inhibitor (-ve) of the F-2,6-BPase catalytic domain, decreasing F-2,6-BP stimulus on kinase domain of PFK-1, stimulating gluconeogenesis.

Ser-32 phosphorylation of the bifunctional enzyme produces a negative charge dependent conformational change of the enzyme, which favors the F-1,6-BPase with a catalytic domain that relates to the family of phosphoglycerate mutases proteins.

If glucose level decreases, glucagon is released into blood, initiating a cAMP signal cascade. In the liver PKA inactivates by phosphorylation the kinase PFK-2 domain (effect lacking in muscle).

If the glucose level increases, the level of F-6-P raises stimulating protein serine/threonine phosphatases. The phosphoprotein phosphatase-1 (PP1) removes the phosphoryl group from the bifunctional protein (PFK-1)/F-1,6-BPase.

The enzyme is activated in the PFK-2 domain and catalyzes the formation of the substrate F-2,6-BP an inhibitor (-ve) of the enzyme F-1,6-BPase, stimulating glycolysis and inhibiting gluconeogenesis.

Cellular Energy Charge

Ion equilibrium allows balancing the responsiveness to hormones of the enzymes AC and insulin receptor tyrosine kinase (IRTK), allowing the adjustments of hormonal responsiveness on target tissues like: fat cells, liver, etc. This provides an integrated network which could support brain requirements during the fight-or-flight response, by overcoming the inhibitory and activatory limits imposed by cellular Energy Charge.

Metabolic intermediates: glucose-6-phosphate, citrate, fructose-1,6-biphosphate, PEP, by decreasing Mg^{2+} were capable to inhibit insulin activation of IRTK activity. [Val^5]angiotensin II increased the kinetic response of IRTK to the MnATP saturation curve [158]. Dysregulation of these pathways commonly occurs during carcinogenesis, possibly due to the prominent role of insulin signaling in mitogenesis. Insulin resistance is a common symptom of diabetes and obesity, resulting from dysregulated energy homeostasis.

The Integrated Mg^{2+}-Dependent Responsiveness to Hormones of AC and IRTK

Signaling by sensor neurons activates the fight-or-flight response of the hypothalamic-pituitary-adrenal (HTPA) axis, increasing the adrenaline output which at the pancreas stimulates secretion of glucagon and inhibits the release of insulin, distorting the pulsatile rhythm of glucose metabolism. This flattens the normal circadian modulation of glucose levels to allow the rate of ATP generation and consumption to exceed the modulatory limits of Energy Charge, and oppose the ATP-requiring regeneration of glycogen and fat reserves. An *in common* signaling results from depletion of chelating metabolites by the opposite tendencies of ionic equilibrium to decrease ATP^{4-}, an inhibitor, and increases free Mg^{2+}, an activator of the hormonal responsiveness of adenylyl cyclases.

Table 10. Partially purified rat liver IRTK preparation. The membrane enzyme was pre-incubated for 15min at 23°C with insulin. Subsequently, was incubated with [γ-32P]ATP, 50 μM, $5x10^3$ cpm/pmol. The kinase activity was measured by the phosphorylation of the exogenous substrate val5-angiotensin II. Activity: nmol/h/mg [165]

Divalent metal	Basal	+insulin	Δ% Insulin-basal
---	0.39	0.22	0
5mM Mg^{2+}	0.5	1.08	+111
5mM Mn^{2+}	9.18	12.18	+33
5mM Ca^{2+}	0.09	0.05	0

The increase in dissociated activatory ion Mg^{2+} vs the decrease of inhibitors CaATP, ATP^{4-}, and Ca^{2+} allows to integrate several tissues in a *in common* AC responsiveness to adrenaline and in fat tissue to additional lipolytic hormones.

This, metabolic network activates the mobilization of glycogen and fats to support liver's gluconeogenesis and its output of glucose. The "n" cooperativity value for Mg^{2+} saturation of the AC increases, because to the saturation by Mg^{2+} activator sums-up with desaturation of ATP^{4-} an inhibitor.

This characterizes a basal AC activity for an "n" value slightly larger than two. The effect of hormones increases "n" to about 4, implicating Mg^{2+}

saturation sites mediating in the membrane, the bonding of AC to adrenaline-receptor and to the activatory mechanism of G-protein.

Energy Charge vs Dynamics of Ionic Mg^{2+} vs ATP^{4-} Equilibrium Balance

The serine/threonine kinase or adenosine 5-monophosphate (AMP) activated protein kinase (AMPK) is a serine/threonine kinase that is activated by upstream kinases LKB1, CaMKKβ and Tak1 among others. AMPK is a heterotrimer consisting of catalytic α-subunits and regulatory β-subunits and γ-subunits.

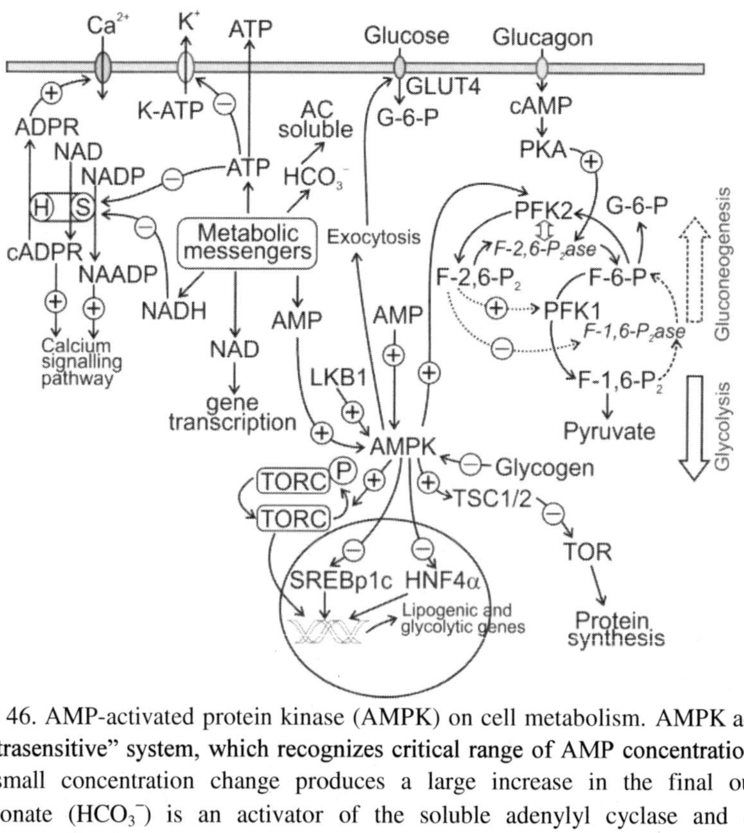

Figure 46. AMP-activated protein kinase (AMPK) on cell metabolism. AMPK acts as an "ultrasensitive" system, which recognizes critical range of AMP concentrations. A very small concentration change produces a large increase in the final output. Bicarbonate (HCO_3^-) is an activator of the soluble adenylyl cyclase and could contribute to the cAMP signaling pathway by activating gene expression of the P2X or P2Y receptors related to ATP activated Ca^{2+} entrance.

AMPK is expressed in a number of tissues, including the liver, brain, and skeletal muscle.

The α-subunit contains the Kinase domain, which transfers a phosphate from ATP to the target protein. The β-subunits and γ-subunits are considered as regulatory components. All three subunits are required for expression of full activity of the AMPK Enzyme Complex, is allosterically regulated by the cellular AMP (adenosine 5-monophosphate)/ATP ratio and by upstream kinases [259].

Full activation of AMPK requires phosphorylation of its activation loop at Thr172 (Threonine-172) of its catalytic α-subunit, by upstream kinases collectively known as AMPKKs (AMPK (AMP-Activated Protein Kinase) Kinases).

AMP stimulates the phosphorylation, both by binding to AMPK and making it a better substrate, and by activating the upstream AMPKKs. AMP binding to AMPK also inhibits dephosphorylation by protein phosphatases.

AMPK activation switches on anabolic pathways and other processes that consume ATP, while switching on catabolic pathways that generate ATP. Examples of the former include acute inhibition of lipid biosynthesis by phosphorylation and inactivation of principal metabolic enzymes [267].

This kinase plays a *Cellular Energy Sensor* role by modulating cellular responses. Catabolic metabolism maintains high levels of ATP, and cells respond to any stress which tends to lower this nucleotide level as an energy challenge. This decreases non-essential ATP-utilizing functions and stimulating available ATP-generating pathways.

AMPK responds to many stress conditions: neuronal necrosis, muscle contraction, metabolic poisoning, sepsis, exercise, hypoxia, inflammation, ischemia, heat shock, etc. AMPK controls many metabolic processes that include increased glucose uptake; increased fatty acid oxidation; decreased fatty acid, glycogen, cholesterol and protein synthesis and induction of mitochondrial biogenesis [268, 269, 270, 271].

In liver can phosphorylate transducer of regulated cAMP response element-binding protein 2 (TORC2). Thus, prevents from acting as a cofactor for the cAMP response element binding protein (CREB) transcription factor, which can activate genes for gluconeogenic enzyme system.

Actived AMPK activates PFK2 (6-Phosphofructo-2-Kinase), which synthesizes F(2,6)P2 (Fructose-2,6-Bisphosphate), the most potent PFK1 (Phosphofructokinase-1) allosteric effector and F(1,6)P2 (Fructose-1,6-Bisphosphate) from F6P (Fructose-6-Phosphate) resulting in the rapid activation of Glycolysis (figure 46).

In summary, AMPK switches the metabolic energy requiring process towards energy conservation. Has rapid effect on processes like glucose entry and glycolysis, an longer-term effects by regulating up DNA transcription coding for glycolytic and lipogenic hormones.

Hypoxia-reoxygenation via the generation of oxidants, likely nitric oxide (NO), AMP, O_2 and $ONOO^-$ increase downstream phosphorylation of PDK-1 (3-phosphoinositide-dependent protein kinase-1) of PI3K.

AMPK becomes a downstream target of PI3K and PDK1, which in turn promotes mitogen-stimulated Akt activation.

Akt phosphorylates eNOS (endothelial nitric oxide synthase) at Ser1177 in hypoxic cells. Direct eNOS phosphorylation by AMPK occurs under conditions of prolonged Hypoxia or ONOO- production or activation by AMP.

AMP inhibits the dephosphorylation and inactivation of AMPK by PP2A (Protein Phosphatase-2A Catalytic Subunit) and PP2C (Protein Phosphatase-2C).

During exercise, AMPK is viewed as a metabolic switch, primarily in light of its ability to inhibit ACC and stimulate Fatty-acid oxidation and Insulin-independent glucose transport in skeletal muscle, in order to avoid problems such as insulin resistance.

Signals for DNA-transcription in the nucleus to expresses a protein and induces some changes such as cell division.

Dysregulation of AMPK has been implicated in a number of metabolic diseases including type 2 diabetes mellitus, obesity development of cancer and neurological conditions [259].

Chronic stress extends the fight-or-flight metabolic conditions, which tend to exhaust ATP. Prevalence of the latter induces tremor in muscle and in brain affects cognitive functions.

It is suggested that free Mg^{2+} integrates a working network according to the hormonal responsiveness of AC and IRTK. The enzymes respond by decreasing the activatory saturation constants for Mg^{2+} and increasing the inhibitory constants for ATP^{4-} and CaATP concentrations.

A small excess of ATP^{4-} over any concentration of MgATP allows to inhibit AC, reducing the rate at which are consumed glycogen and fat-stores. This adapts the kinetic response of AC to extend to a new limit, the metabolic requirements of the fight-or-flight reaction.

AMPK is activated by a transient decreases in the cellular energy charge leading to AMP-driven phosphorylation by oxidative pathways generating ATP. Its accumulation signals a reverse of directional tendency stimulating ATP consuming-biosynthetic pathways.

EC could be given a role in ionic equilibrium by its relationship to the control of the availability of free divalent metals. The latter, related to the stronger binding constants of ATP^{4-} for Mg^{2+} and Ca^{2+} than for ADP^{3-} and AMP^{2-}.

Opposite metabolic senses could not tend to operate at the same time. Under the proposed formulation of EC the relation between pathways could be considered according to the dominant pathway with a turn-off effect in the biosynthetic enzymes by AMPK.

This involves oscillatory synchronization of oxidative vs biosynthetic pathways with narrow circadian rhythms peaks, and theoretically could not take into account the role of adrenaline for the stress-amplification of the use of energy stores, during flight-or-fight.

The role of AC at different tissues allows this to integrate into a network by the *in common* dependence of the ionic Mg^{2+}.

An increase in free Mg^{2+}, exceeding MgATP formation could increase the responsiveness of the enzymes in the membranes, IRTK and AC to insulin and adrenaline respectively.

Chelating metabolites are inhibitory on the discussed membrane enzyme as shows for $citrate^{3-}$, which function as a chelators of free Mg^{2+}, preventing its activatory effect and inhibiting the membrane enzyme by the release of ATP^{4-} from MgATP.

Formulation of Atkinson's Energy Charge (EC) =[ATP]+ 1/2[ADP]/([ATP]+[ADP]+ [AMP]) [272] can have a value ranging from 0 (all AMP) to 1 (all ATP). The values regulate the activation vs deactivation of endogenous cellular enzymes. If approaches the value of 1 activates ATP-dependent-biosynthetic pathways, forming energy stores such as synthesis of glycogen, lipids, etc. And at the same time prevent activation of oxidative pathways. Values of 0.5 and lower indicate an excess of adenosine monophosphate (AMP), which modulates to increase the activity of the oxidative pathways to generate ATP for energy-consumption processes. ATP-activatory and AMP-activatory signaling are mutually exclusive antagonists.

A dynamics of ionic-metabolic coupling (IMC) by changing the relative concentration of $\Delta[Mg^{2+}]/\Delta[ATP^{4-}]$, may be extended to include the presence of Ca^{2+} $\Delta[Mg^{2+}]/\Delta[ATP^{4-}]$ $[ATPCa][Ca^{2+}]$. An increment of a metabolic rate could persist even at low Mg^{2+}, because the stressing hormone adrenaline, increases AC affinity for Mg^{2+} and decreases that for ATP^{4-}. The adrenaline-mediated increment of the liver output of glucose could persist even in the presence of accumulated chelating molecules like ATP^{4-}, if do allow an excess of free Mg^{2+}.

Ionic control allows that tissues under hormonal signaling can collaborate in support of the metabolic energy flow.

A decreasing in cellular MgATP increases the ion Mg^{2+} required for adrenaline stimulation of adenylyl cyclase at the tissue network of liver and fat cells, which are connected by the energy requirements of the fight-or-flight response.

The resulting glycogenolysis and glucose oxidation in liver combines with the release of fatty acids from fats and amino acids from the glucagon stimulated proteolysis to generate glucose. This molecule supplies to blood the energy to support the fight-or-flight response. Overstimulation by adrenaline and cortisol produces exhaustion of metabolic stores (glycogen and fats). *Evolution* favors the brain dominance through the absence of a negative feedback of adrenaline because this molecules lacking the capability to cross the BBB.

Adrenaline decreasing insulin-output and distorting circadian rhythms extends cellular starvation at many body level tissues but not to the brain itself supported by the simultaneous turn-on of gluconeogenesis in liver supplying glucose to blood in support of brain vital functions, but affecting emotional-linked behavior, characterized as psychosomatic dysfunctions.

The noradrenaline-activated-AC at the HTPA axis through an ionic-metabolic coupling controls the responsiveness to adrenaline of AC in metabolic network of tissues in the soma/body supporting brain. The lack of an adrenaline negative feedback on the HTPA axis, results in the psyches inability to exercise in the soma homeostatic turn-off controls that usually preserve excessive use of fat and glycogen stores.

The responsiveness to hormonal and Mg^{2+}-mediated activation of AC sustains a steady state generation and consumption of ATP, which could support for a longer time and stress event commanding unusual expenditures of energy. The stress persistent-increment of adrenaline could overcome its rate of destruction and by lacking a negative feedback, could maintain in an overstimulated state the function of the HTPA-axis.

Integration of the ionic and metabolic cellular controls of the insulin receptor tyrosine-kinase (IRTK) and adenylyl cyclase activities (AC). An increment in the concentration of chelating metabolites produces the release of free ATP^{4-}, a negative modulator of AC and IRTK. Magnification effects by simultaneous decrease of the chelated substrates: concentrations of MnATP and MgATP, and of the activatory modulators the free ions: Mn^{2+} and Mg^{2+}.

The endogenous cellular Mg^{2+} responsiveness of AC receptor in the membrane to determine which tissues could be stimulated by a hormone present in blood that by a phosphorylation cascade articulates synchronization of activation-desactivation of the cellular endogenous enzymes within a network.

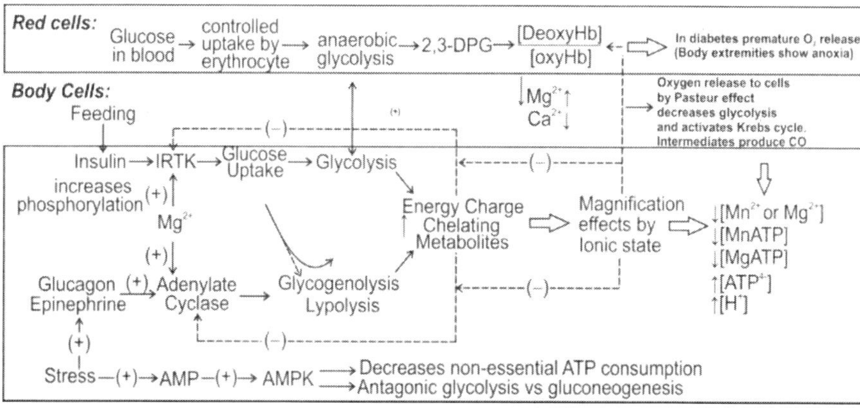

Scheme 3. O_2 uptake by the Pasteur Effect decreases glycolysis and activates the Krebs cycle. A decrease in chelating metabolism favors the release of Mg^{2+}. AC and IRTK responsiveness to hormone is increase by the relative concentration of the activators Mg^{2+}/Mn^{2+}, versus the inhibitors Ca^{2+}, ATP^{4-} and CaATP.

Chapter 9

Responses to Stress and Associated Dysfunctions

Medical research of 1895 lacked the means to elucidate brain-body molecular interrelationships. Hence, most of the advances in the treatment of mental diseases rested in Freud and his followers' development of a theory assigning to the unconscious mind, the etiology of mental diseases.

Therapies for mental health based in causation differ widely. Psychoanalysis assigns to the unconscious mind, the etiology of mental diseases. Ethology emphasizes evolutionary, adaptive behavior. Brain anatomy differentiated cells structure and functions of the component cells neurons, astrocytes, oligodendrocytes, establishing multiple area correlation and localization of sensor regions, etc. Electrophysiology characterized the electrogenic membranes of neurons correlated by behavior biology to neuronal action-response specific parameters, biochemistry added the thermodynamics and metabolic connections involved. Many scientific advances have consolidated the parameters, all present in the cause and effect relationships, which partially describe the normal mind-brain function and could be involved in mental dysfunctions.

The study of evolutionary and adaptive response of human behavior by Ethology resulted in causation etiology, emphasizing the role of cultural stress on mental health related diseases [273]. Therapies for mental health differ widely if based in causation. The human cultural environment offers many examples of useful assistances to prevent stressed individuals to reach the acute or chronic stages of psychotic related diseases.

It is widely accepted that the modern life style induces the stress-related fight-or-flight response. Adaptation became an individual mode to reach an allostatic state. The continuous/chronic presence of stress could be accepted by many individuals in order to obtain psychological and social well-being. A short-term activation of the hypothalamic-pituitary-adrenal (HTPA) axis involves that energy substrates could be metabolically canalized to meet the increased organismal demands.

Catecholamines are the primary stress hormones because noradrenaline (NA) activation of the HTPA axis increases adrenaline and cortisol levels. The latter is also referred as a primary, because potentiate the role of a high adrenaline level by jointly stimulating glycogenolysis, white adipose tissue lipolysis and the inhibition of triglyceride synthesis.

Metabolite accumulation could decrease the concentration of free divalent metals, required for the binding of proteins into their membranes [274, 275]. Accordingly, protein apoptosis rates through in excess of ATP^{4-}, extracting the divalent metal that participate in a protein folding and its binding to a membrane, could provide a mechanism for an established characterization of feeding restrictions over life expectancy.

To maintain for a prolonged time a flow of energy the fight-or-flight metabolic response overcomes cellular energy charge signaling turn-on and -off pathways. The metabolism shifts from its regulation by activatory vs inhibitory metabolites to a hormone-dependent control over enzyme activity. The latter usually mediated by phosphorylation and dephosphorylation, interconverting the active vs the inactive forms of an enzyme.

NA activates the hypothalamus-pituitary adrenal axis to stimulate the secretion of corticotropin-releasing factor (CRF) from the hypothalamus, which induces adrenocorticotropic hormone (ACTH) release from pituitary stimulating at the adrenal glands cortisol synthesis and adrenaline release. NA by feedback inhibits its production at LC, and CRF feedback to inhibit its production, but stimulating feedback over LC to increase NA [276]. This effect on the hypothalamic-pituitary-adrenal (HTPA) axis could evolve into chronic stress, and cannot be turn-off by the increase in adrenaline in blood, which is prevented from a negative feedback in brain because, is not able to cross the blood-brain barrier (BBB).

The adrenaline-mediated stress became present at the level of the unconscious, where most of the work of the mind gets done. The absence of a physiological pathway for direct mind control of the deleterious effects of adrenaline at the body itself could be characterizable as psychosomatic trauma or disorders.

Stress is a response to adrenaline and cortisol characterized by its participation on the emotional enhancement of memories, which is suppress by amygdala damage. Excessive or cortisol prolonged stress hurts memory storage [277, 278].

The Freudian methodology allows medical assistance of the affected individual in order to decrease is danger-perception and therefore provides means to turn-off the fight-or-flight response, major psychological sources of psychological stress.

The mutual interaction between mind and the molecular level reveals the convenience for the mental stress problematic to be derived at the interdisciplinary convergence of psychological, psychiatric, neurology and molecular biology methodologies. Progress of unrestricted fight-or-flight shifts, by increasing the adrenaline-dependent depletion of the glycogen and fatty acid the stores and preventing insulin control affects the circadian rhythms, which promote metabolic balances and homeostasis, affecting support of mental processes. The further NA-stimulus of the HTPA-axis leads to increase the level of cortisol to obtain free amino acids from protein breakdown to maintain uncontrolled gluconeogenesis. This, is required to support brain, but with the cost of consuming proteins from other body organs, including heart.

Stress Mediated Dysfunctions

Stress mediated by increasing adrenaline secretion, usually shows a pattern of hyperfunction follow by hypofunction, which when the stressed organ is brain it could be associated with a symptomatology of persistent anxiety followed by depression [96].

Under prolonged stress, the synthesis of new AC, may not match the rate of noradrenaline-dependent inactivation. The viability of functional synapses may be controlled by the speed of synthesis of the proteins, which make-up the hormonal receptor sites and/or AC-itself, versus the rate of inactivation or destruction of AC. This generalized molecular modeling for the evolution of psychosomatic diseases [96] could be tested with regard to more specific physiological conditions and permit prognosis of dysfunction, allowing preventing treatments. Already medical applications have been developed, based in the use of beta-blockers to interfere with a stressful condition, attenuating the memory of a traumatic event.

Acute Stress

Noradrenaline (NA) injected into the vein acts on the α type of adrenoceptors. Its effect augments body performance by increasing heart rate, blood pressure, widening of pupils and the air passages of the lungs. The α receptors are found in muscle inside the walls of the peripheral blood vessels to contract and narrow. In the extremities by redirecting blood to essential organs like heart and brain, NA can produces greater resistance for the heart beating and increases blood pressure. Injected NA could be used to restore blood pressure in emergency situations [18, 19, 20].

The risk of raising too high the blood pressure its presence and NA may affect the heart rhythm. Therefore under non-emergency situations: linezolid, beta-blockers such as propranolol, MAOI (monoamine oxidase inhibitor) antidepressants such as phenelzine, rasagiline or selegiline for Parkinson's disease, general anaesthetics, and tricyclic antidepressants such as amitriptyline.

NA can reduce the blood supply to the placenta, but it is poorly absorbed from the gut and therefore do not pass into breast milk.

Acute stress disorder is a psychological condition arising in response to a terrifying or traumatic event like psychological shock, mental shock, or simply shock.

NA activates the β-adrenergic receptors (β-ARs) to increase the brain cytokine IL-1β production in the amygdala with concomitant increase in the NA turnover in the prefrontal cortex, hypothalamus, and amygdala.

Acute stress via NA activates the β-adrenergic receptors (β-ARs) to increase the brain cytokine IL-1β production in the amygdala with concomitant increase in the NA turnover in the prefrontal cortex, hypothalamus, and amygdala.

Acute-panic could be related to the corpus coeruleus axons of a large discharge of NA that at the postsynaptic cleft located AC, which if not promptly reuptaked could result in the inactivatory overexposure of AC to NA. Thus, if the inactivatory rate exceeds the synthetizing rate of AC, could develop into a symptomatology of anxiety followed by depression. At the BBB oxyHb delivers O_2 and Mg^{2+}, the ion is required to stimulate responsiveness to NA of postsynaptic AC.

Investigators have reported that *in vivo* stimulation of brain induced a glucose uptake and metabolization, ahead of the increase in O_2 consumption. In neurons the insulin-responsive transporter GLUT3-glucose complex, crosses the BBB. This allows inferring, not only nutritional but also dysfunctional signaling, by the flattening of glucose-dependent circadian rhythms.

Anxiety and Depression

Serotonin functions include the regulation of mood and many behaviors appetite, hunger, sleep, sexual response, impulse control involving emotions and judgment and anger, depression, anxiety and perception. High levels of serotonin characterize tendency overstimulated individual, under pressure for over-exacting completion of tasks and those with insomnia. Overstimulation of serotonin induces anxiety and reducing calories intake may induce some calming effects.

The activity of the neurotransmitter-enzyme complex correlates to a reinforcement over the basal activity one elicited by free Mg^{2+}- or Mn^{2+}-dependent activation of the basal state of the enzyme. However, prolonged effects may elicit a central nervous system (CNS) excited or anxious state which prolonged results in a depressive state [279]. The latter, could be related to memory deficiencies so emotional potentiation of memory capability collapses limited under this stress condition.

Milder but prolonged panic attacks by over- stimulating AC could lead to diminish the enzyme activity, effect characterizable within a biphasic pattern of anxiety followed by depression.

Stress is a major risk factor for the development of mood and anxiety disorders.

Depressed patients are in state of low mood and report sadness, anxiousness, lack of sense of well-being [280], constantly worried, guilty, irritable hurt, restless. Their feelings include perception to be alone, loss of interest in every day affaires have problems concentrating, making decisions, aversion to activity, and loss of appetite or overeating, digestive problems, reduce energy, excessive sleeping aches, pains fatigue. Sometime contemplate to commit suicide.

"Endocannabinoids" are all eicosanoids, the synthetizing and degrading enzymes fatty acid amide hydrolase or monoacylglycerol lipase.

Neuromodulatory lipids and their receptors constitute the endocannabinoid system involved in pain-sensation, appetite, mood and memory.

The endogenous arachidonate-based-lipids, 2-arachidonoylglycerol (2-AG) and anandamide (*N*-arachidonoylethanolamide, AEA) are ligands for the cannabinoid receptors [281]. These are the CB_1 and CB_2, two G-protein-coupled receptors, which are located in the central and peripheral nervous systems.

Pharmacological augmentation of central endogenous cannabinoid (eCB) signaling may be an effective therapeutic strategy to mitigate the adverse behavioral and physiological consequences of stress [282].

Cannabinoid type-1 (CB1) receptors are widely distributed in the central nervous system and are involved in the modulation of emotion, stress andhabituation responses, behaviours which are thought to be dysregulated in human psychiatric disorders [283].

On targeting medication to control the intracellular signaling pathways and molecular switches for efflux transporter regulation allow fine-tuning by turning off for controlled periods. There are clinical report that detects reduced serum zinc and magnesium in depression, which can be normalized by diet, containing these divalent metals, before resorting to antidepressant treatment. Blocking the mechanism of reabsorption may lead to an increase in the availability of serotonin neurotransmitter.

Euglena, a genus of single-celled flagellate protists, cultured under organic phosphate deprivation in addition to the presence of the constitutive phosphatase, activates the genetic expression of a separated enzyme. Thus, under environmental stress the unicellular organism responds producing an induced phosphatase. This one is about four times more heat-labile than the constitutive phosphatase, which is indicated a very primitive evolutionary adaptation, which associates the inductive and this reversion mechanisms [284]. Antidepressant drugs work by helping to restore the brain's chemical balance. Drug therapy for depression may take up to 1 month or more to become fully effective. This time-dependence could indicate participation of the reversion mechanism. The complex formed between adrenalines and adenylyl cyclase decreases the stability of the enzyme itself. This indicates that adaptive mechanisms similar in nature could have been associated to the turn-off of environmental stress by higher species along the evolutionary scale.

Chronic Stress

Cental nervous system (CNS) sympathetic pathways are activated for enhanced alertness and focused attention, but feeding, sleep, sexual drive as well as other vegetative functions became decreased. Peripherally, the humoral and neural systems support the most pressing requirements by elevating heart rate, blood pressure, respiratory rate, and gluconeogenesis.

Cognitive emotional or like inflammatory a non-cognitive stressor triggers processes that tend to adapt, however, the emotional pressure for a prolonged period overcomes the feeling of control. Presynaptic noradrenaline releasing neurons act on postsynaptic noradrenaline-dependent neurons at the HTPA axis, activating the adrenals cortex response with release of corticosteroids and the medulla by releasing adrenaline. These stress hormones help to cope over some time with the stressor, but long-term exposure to stress maintains constant a high level of the hormones physiologically required to made available energy stores to confront a fight-or-flight situation. Thus, adrenaline maintains high blood pressure and cortisol by utilizing amino acid from degrading heart or muscle tissue, inhibition of growth, suppression of the immune system [285] lead to mental dysfunction.

Parkinson

Parkinson's disease is characterized by loss of dopaminergic neurons; treatment usually includes DOPA, likely to partially alleviate the symptoms leading into the idea of an all-inclusive DOPA-reward mechanism. The cerebello-thalamo-cortical (CbTC) interactions linked to metabolic dysfunctions in Parkinson's patients also characterize the disease.

Adrenaline produced by the adrenal medulla chromaffin cells could have a direct effect on increasing corporal energy mobilization for muscle by increasing heart rate and blood pressure but overtime exhausts muscle glycogen and fat reserves. Tremors indicate insufficient ATP for the sliding mechanism of muscle, which could occur after an excess of muscle exercised up to exhaustion. The effect under physiological conditions can be reversed by resting, but under stress could become a symptom of metabolic exhaustion.

Pharmacological agents targeting dopaminergic/serotoninergic neurotransmission have been clinically used to manage several neurological and psychiatric disorders including Parkinson's disease, schizophrenia, bipolar disorder, depression, attention deficit and hyperactivity disorder (ADHD), and addiction.

Suprachiasmatic nucleus (SCN) neurons remain steady in the circadian expression of intracellular levels of mammalian circadian clock genes (Clk).

Effect not observed in the hippocampus or brainstem. Repeated stressor exposure decreases the β-AR expression at the onset of depression [286].

Persistent adrenaline-dependent inhibition of insulin secretion interferes with the possibility to maintain a circadian rhythm, controlling availability of glucose to muscle, and brain. Parkinson and schizophrenia share on muscle-tremor symptomatology which relates to depletion of the ATP required for muscle sliding. Insulin allows glucose to feed muscle, has been clinically observed to decrease muscle tremor.

The overexposure to adrenaline has an inactivatory effect on the adrenaline-AC complex, which is considerably less stable, than the complexes of AC with either glucagon or adrenocorticotropic hormone (ACTH). The role of BBB extends to prevent that the adrenaline-mediated stress response, could trigger a non-localized, over-excited activity to all brain areas, rather than the ones directly related to an emotional-signaling. Evolution separated stress exerted on the body from the one on the mind/brain.

Treatment by DOPA (3,4-dihydroxyphenylalanine) was likely to partially alleviate the symptoms. Reserpine, one of the first tranquilizers, inhibits the dopamine-transport to the interior of the vesicles.

Tremor-related patterns (PDTP) and loss of dopaminergic neurons characterize Parkinson's disease. Dopamine (DOPA) may alleviate the symptoms, by a differential effect on glucose-induced insulin secretion. Noradrenaline at 10^{-1} mM has been show to inhibit the dopamine uptake as well as its stimulatory effect at 10^{-5} mM on glucose-induced insulin secretion. Cocaine blocks the reuptake of catecholamines, prolonging their action.

At the pancreatic β cells the expressed DOPA D2 receptor mediates inhibition of insulin secretion [249]. The pharmacological agents that block the dopamine receptors have been used for the treatment of schizophrenic patients, but in occasion induce tremors similarly to Parkinson's disease.

Alzheimer

Efflux transporters are affected by and likely contribute to disease pathology of central nervous system (CNS) disorders that are accompanied by inflammation, oxidative stress and neurotransmitter release, and that include cancer, epilepsy and Alzheimer's.

Activation of xenobiotic-sensing nuclear receptors like the aryl hydrocarbon receptor (AhR), the glucocorticoid receptor, the pregnane xenobiotic receptor (PXR) and the constitutive androstane receptor (CAR) regulate transporter expression.

The ligands for the insulin-signaling pathways, hormone insulin and insulin-like growth factors I (IGF1) and II (IGF2), bind to the insulin and IGF1 receptors, with downstream signaling by the MAPK and PI3K pathways to activate target gene expression. Insulin resistance is a common symptom of diabetes and obesity, resulting from dysregulated energy homeostasis.

Also Alzheimer's patients show insulin resistance associated with the cognitive decline. This relates to somatomedin C (IGF-1) resistance, and the insulin receptor substrate 1 (IRS-1) (adapter protein for signaling, in humans encoded by the IRS-1 gene) dysregulation [287].

Decreased blood-brain barrier P-glycoprotein (Pgp) function has been shown in Alzheimer's disease (AD) patients using positron emission; activation of PXR restores brain endothelial which resulted in enhanced amyloidβclearance from the brain [288].

Memantine (5 Methylcobalamin (methyl-B12)) approved by the US Food and Drug Administration because it has a direct impact on the individual's ability to function in real life. It is a cerebral protector which acts over the glutaminergic system and reduces the chemicals in brain with can contribute to the Alzheimer symptoms and has been shown improving cognition in bipolar disorders.

CREB may possess therapeutic potential for patients that have Alzheimer's disease.

Schizophrenia

The 23 genes to undergo frequent genomic alterations in schizophrenia, typically present in young adulthood, thought processes break down, emotional responsiveness becomes poor, and significant social or occupational dysfunction often occurs. Mental health professionals diagnose schizophrenia based on observed behavior and reported experiences and treat it mainly with antipsychotic medications, which can have complications, and psychological and social support, which may or may not be wholly effective. Several de novo genomic rearrangements, including copy number variations (CNV), have been identified at specific gene loci in patients with schizophrenia. The genes on the array encode anchoring/scaffolding proteins, cell adhesion molecules, cytoskeletal components, ion transporters, metabolic enzymes, neurotransmitter receptors, and transcription factors. These proteins regulate processes such as axon guidance, signaling (via hormones, nerve growth factors, and neurotransmitters), nervous system development, neurogenesis, synaptic plasticity and synaptic transmission.

Schizophrenia is frequently accompanied by metabolic syndromes like type 2 diabetes, dyslipidaemia, etc. Post-mortem brain tissue shows alterations in glucose metabolism and insulin signaling pathways.

Over-Feeding

Currently, the most important factor known to limit individual lifespan within given specie for example: rats, appears to be over-feeding. This condition could lead to the prevalence of high ATP^{4-} and other chelators.

Hypoglycemia could install defective glucose-signaling pathways obstructing some of the genetic expressions pertinent to brain function [162].

Hypoglycemia leads to cerebral injury, but increased the expressions of glucose transporters (GLUT1, GLUT3) and nuclear factor kappa-light-chain-enhancer of activated B cells (NF-κB) and 4-Hydroxynonenal (HNE) which are suppressed by Cr-chelates (CrHis). The latter enhanced expression of neural cell adhesion molecule (NCAM), growth associated protein 43 (GAP-43) and Nuclear factor (erythroid-derived 2)-like 2 (Nrf2) a transcription factor. This indicated a chromium protective effect and its regenerative potential for the modulation of neuroplasticity and nuclear transcription proteins [289].

It is suggested that the chromium-chelation of His-groups of proteins configures complexes, which are more stable than the ones characterizing Mg^{2+}-, Zn^{2+}- and Mn^{2+}-protein complexes. Hence, a lower tendency for metabolite-dependent turnover of the chromium complexes results into consolidating a genetic expression forms similar to constitutive.

A mechanism for apoptosis proposes a selective differentiation between two folding conformations [290] or that of denatured protein vs its original configuration. Protein denaturing involves precipitation by the loss of solvation water.

Partial loss of hydration shells inactivates ATPase [291] by preventing the conformational chain required for catalytic turnover.

Bilateral Electrocoagulation of Adrenal Medulla (BEAM) a Surgical Treatment for Stress-Induced Metabolic Dysfunction

Chronic-stress-treatments by drugs capable of blocking the effect of hyper secretion of adrenaline and cortisol are presently under development by many laboratories.

Experimental animals (dogs) treated with amphetamines to induce psychotic symptoms were used to test the effect of surgical elimination of the production of adrenaline, and was observed that psychosis and aggressiveness disappeared.

Mackliff's technic was denominated a Bilateral Electrocoagulation of Adrenal Medulla (BEAM). The BEAM treatment to 40 patients with chronic schizophrenic and Parkinson diseases shows that adrenaline disappears and noradrenaline was diminished 20% in the blood stream. The procedure returns cortisol to normal levels. BEAM suppression of adrenaline secretion allows the cortisol mediated-negative feedback on anterior pituitary and hypothalamus, and restores glucoregulation [292].

The circadian effect on metabolic processing of glucose in brain allows that the Parkinson increased cortisol level, in the absence of adrenaline could manifest a negative feedback on the axis and decreases tremor.

The results show a better response to stress, solving the axis fatigue syndrome. The shifting to an adrenaline lacking metabolism could be explained by restoring a hormonal responsiveness of AC. BEAM suppresses further deterioration in the body of a functional AC.

Evolution favored to preserve a fraction of AC from adrenaline overexposure, by conferring greater stability adrenocorticotropic hormone (ACTH)-AC and glucagon-AC complexes, which are more stable than adrenaline-AC, allowing metabolism in which the resultant cyclic adenosine monophosphate (cAMP) acting as a second messenger could control fat stores, which support liver gluconeogenesis and mobilization of glycogen stores to support the liver output of glucose.

In-vitro overexposure to catecholamines of adenylyl cyclase of fat cell membranes let to the loss of its catalytic activity by prolonged exposure. BEAM restores a pattern of glucose availability to support a circadian rhythm adequate to the motor features affected by these disorders.

Muscle tremor could be physiological for an exhausted runner, because muscle works by a mechanism of myosin vs actin, contracting by inflow of

Ca^{2+} and their sliding by MgATP. However, could be associated to a chronic stress exhausting ATP in muscle.

BEAM eliminates the effect of adrenaline inhibiting insulin secretion. In similitude to the use of the radioisotope of iodine (I-131) for hyperthyroid control, a reduction of the secretion of adrenaline could be obtained by radioisotope Sr-89-silicon inserts into the adrenal medulla [293], calculations shows highly focused tissue destruction by radiation [294].

Stress related diseases manifest an abnormally flat circadian cycle, because the negative feedback by cortisol in HTPA-axis, could not overcome the over stimulation by adrenaline of the sympathetic system, and the loss of glucoregulation affects the brain. The insulin shock shows that the brain state of consciousness is glucose-dependent [295].

The removal of adrenaline could favor partial restoration of some metabolic controls. Hence, a decrease in the adrenaline overstimulated state of the HTPA axis favors effective feedbacks by ACTH and cortisol. A decrease of cortisol in blood could turn to normalize gluconeogenesis and stimulate in pulsatile manner the release of glucagon and insulin discharges. A partially improvement of circadian rhythms could also result from the annulment of adrenaline interference with insulin secretion [292]. Thus, a new circadian rhythm for glucose may appear based in the sequence of pulsatile peaks for the release of cortisol vs insulin.

Chapter 10

Conclusion

Isolation and Characterization of Noradrenaline-Activated Adenylyl Cyclase of Brain

The experimental studies were directed:

I) To elucidate the kinetic properties of AC relevant to modeling mechanism:
   a) for noradrenaline-dependent emotional eliciting and reinforcement of AC-dependent long-term memory,
   b) for continuous memory processing by activation of the basal AC activity by free divalent metals;

II) Characterization of hemoglobin red cell-system as a sensor of glucose levels in blood, and responsiveness for O_2-release matching glucose uptake

III) To detect a metabolic feedbacks of a tissue network, capable to shift the rates of O_2 and glucose consumption, between hormonal activated and the basal memory processing stage. The matching of the rates of brain glucose consumption and tissue oxygenation requires *in common* molecular signaling. This one should operate within a physiological network conferring to the red cell-Hb system a sensorial role of glucose levels by its uptake into the erythrocyte as the source of 2,3-DPG generation capable to decrease Hb affinity for O_2 [1, 2, 4].

LC Presynaptic Release of Noradrenaline Jointly with Mg^{2+} Released from Red Cell Activates the Postsynaptic Adenylyl Cyclase

To examine networking events by their functional connectivity requires to evidence the chronology of the contributions to the development of the overall system. Locus coeruleus (LC) neuronal axons release noradrenaline (NA) into receptor neurons. At the blood-brain barrier the glucose in plasma can transvers the barrier by diffusion and forms a glucose-GLUT3 complex which transport across the membrane.

It may not only signal availability of glucose but association-dissociation, involving gain and loss of H-bonds, confers a vibrational pattern into the membrane, which could modulate an action potential. Glucose activates the glucose sensorial neurons at the hypothalamus which is sensitive to glucose levels.

The impulses from corpus coeruleus and the synapses from other activate neurons are integrated in the cell body of a receiving neuron. Thus, results in widespread emotional connectivity within neuronal circuits of diverse brain areas [162]. Activation of concatenated neurons could repeat sequences of action potentials, which could code differentiable stimulus [296, 297].

Mg^{2+} in opposition to Ca^{2+} facilitates crossing of the BBB and reverse the inhibitory effect of Ca^{2+} in the responsiveness of adenylyl cyclase to NA. The choroid plexus consists of modified ependymal cells structured in the ventricles of the brain where cerebrospinal fluid (CSF) is produced. In the brain 10^{11} neurons are connected each one with about 4μm of surrounding arterioles, allowing individualized rapid transport of Mg^{2+} and Ca^{2+} ions across the Blood-CSF-Barrier.

Metabotropic and ionotropic receptors are related to synaptic plasticity. The change in the number of ionotropic receptors at the postsynaptic neuron induces activation by increasing or depression by decreasing.

The ionotropic glutamate receptor/N-methyl-D-aspartate receptor (NMDAR) participates in the control of synaptic plasticity and memory function [298], maintaining the Mg^{2+} vs Ca^{2+} responses in opposite activated state and overall fluctuations in the activatory vs inhibitory processes. The metabotropic receptor could modulate through second messengers synaptic plasticity by activating protein synthesis at the postsynaptic cell.

Glutamate receptors are present in the glial cells and may play a role by increasing extrution of water exhausted by H-bond decreased clustering. It is suggested that this process participates of the night-dependent contraction of brain.

Presynaptic terminals in hippocampal cultures during uncorrelated activity show a transient enhancement of synaptic plasticity by reduced Ca^{2+} flux [299], a more permanent effect was obtain via adjustment of the voltage-dependent Mg^{2+} block of the NMDAR.

The Biochemical Level of Emotional Intelligence

Biochemistry progress is indebted with validation of physicochemical parameters, which could allow encoding brain states, and their retrieval, at the molecular level.

Experimental findings and theoretical ones [95, 300, 301, 302] were applied to a model of energy transfer through the turnover of hydration shells of proteins and ions [161, 272, 303]. This could be expected to contribute to the understanding modulation of coupling in mechanism for molecular signal-electrogenic transduction.

The nature of information decoding mechanism may relate to the coupling of energy transduction processes through the modulation of the quaternary structures of proteins. The short-term memory (STM) mechanism could be based in the fluctuation between two states of hydration/solvation of coding proteins [125, 126, 304, 305, 306, 307, 308].

In this integrative mechanism, noradrenaline (NA) and nitric oxide increase arteriole diameter. At the homeostatic pH=7.4 of brain the plasma glucose is uptake by the erythrocyte and generates 2,3-DPG. The latter, by binding oxyHb produces deoxyHb and released O_2 and Mg^{2+}.

The Mg^{2+} ion at the vascular endothelial cells of BBB reverses the Ca^{2+} tightening at these cells junctions and therefore increases permeability. This allows that glucose could be discharged into cerebrospinal fluid (CSF) and metabolized by neurons before O_2 [297]. NR2A subunit protein in the neuronal surfaces controls influx and efflux of Mg^{2+} and Ca^{2+}.

Hb could act as the divalent metal carrier by the oxy- to deoxy-Hb turnover allows the protein to change its conformation with the loss of 60 water molecules, and breaking of their corresponding H-bonds to decrease affinity for the ion Mg^{2+}. The LC axons release from stores Ca^{2+} which acts for NA vesicles discharge at the presynaptic space.

Mg^{2+} reaches the cleft activating the postsynaptically located NA-stimulated AC and thus reverses the inhibitory effect of Ca^{2+} on AC. The enzyme could function in a pulsatile manner because for NA in the presence of Mg^{2+} shows positive cooperativity with n=4, whereas the inhibition by Ca^{2+} shows negative cooperativity of n=0.6 [1, 2, 4].

Action potentials may propagate according to synaptic integration but conditioned by the crossing of the BBB by Mg^{2+} displacing Ca^{2+} and the glucose-carrier complexes signaling as a reward response to neurons increasing synaptic connectivity.

The LC axons by releasing NA mediate long-ranging Ca^{2+}-signals by activating astrocytes α_1-adrenergic receptors [309]. The cerebrospinal fluid carries the H-bond exhausted water Na^+ and K^+ rapidly along glial channels on the outside of arteries that at the other side of tissues enters into the veins that eventually drains into the body's lymphatic system.

160ml of cerebrospinal fluid (CSF) is recycled 3.7 times per day in order to clean out metabolites and toxins like β amyloid. Hence the choroid plexus must produce about 500 milliliters of CSF daily (or 20.83 ml per hour).

The turnover of hydration shells of ions, receptors, enzymes, transporters and channel carriers require H-bond losses paid for by surrounding water clusters, which could be completed depleted of H-bonds if eventually exhales as vapor. An ionic kosmotropes Mg^{2+} and Na^+ have stronger structuring interactions with water, than the latter molecules interacting in a water cluster structure. An ionic chaotrope like K^+ exhibits weaker interactions with water and lower interference on the H-bonds pattern of surrounding water. These interactions allow thermodynamically metastable phases, on internal chemical bonding of molecules surrounded by water.

The thermodynamics involves interaction of Mg^{2+} with channels proteins which by ions influx-dependent modification, fluctuate from a hydration state inactive to one active.

The time for neuronal firing may respond to Na^+/K^+-exchanges, by signaling through changes on the hydration shells that the sizes of the ion adequate their hydration shell structure of Na^+ and K^+ to fit into the corresponding channel and charging positively the membrane along the action potential path.

The mitochondrial generation of ATP becomes overcome releasing Mg^{2+} during the rapid Na^+ pump operation and therefore activates AC responsiveness to NA.

Functional conformational change of proteins require turnover to be driven by the breaking of H-bonds, in between water molecules.

The Mg^{2+}-dependent bonding of a specific receptor on several neurons could assign specificity to a circuit and by incorporating other receptors into the membrane of other neurons, could form differentiable connectivity patterns.

Human muscle diseases have been reported as linked to ion channels disorders, by dysfunctions in the action potentials, signaled from the neuromuscular junction [202, 310]. Hyperexcitability can lead to unwanted and persistent contraction, hypoexcitability is reduced [200, 311]. The β-adrenergic/cyclic adenosine monophosphate (cAMP) mechanism in long-term memory (LTM) shows that responsiveness to β-adrenergic receptor of the tissue network supporting brain metabolism manifest an *in common* dependence on free Mg^{2+} [2, 4, 6, 8, 14].

The forming of cAMP-response element-binding proteins (CREB) activating the CRE region for DNA transcription.

Genes whose transcription is regulated by CREB include: c-fos, the neurotrophin BDNF (brain-derived neurotrophic factor), tyrosine hydroxylase and many neuropeptides like corticotropin-releasing hormone, somatostatin, enkephalin, and inducible nerve growth factor (VGF) that regulates energy homeostasis and synaptic plasticity [312].

CREB has a role in neuronal plasticity and long-term memory formation in the brain. CREB has been shown to be integral in the formation of spatial memory.

The generated polypeptide molecules could act by binding through Mg^{2+} to receptors within the membrane. These could become integrated within a neuronal circuit by an *in common* vibrational resonance.

A mechanism separating emotional-LTM from STM also called working memory may not require totally independent pathways. Both could operate from CREB mediate process of accumulation of transduced information into peptides. The latter, could be selectively encoded within the membrane. Natural selection may adjudicate lasting neuronal patterns, out of nurturing experiences, along the life learning period, which characterizes each animal species. This could be exampled by nurturing, sex, etc., could activate CREB-dependent induce genes involved in the oxytocin peptide synthesis. Working memory could be expressed because emotional conditions lead to stimulate constitutive gene expression of oxytocin recognition receptors. The increment of these ones at the spines and/or dendrites could condition a rapid response to the stimulated release of Mg^{2+}. The Mg^{2+} activatory binding of the oxytocin-occupied receptor to the membrane sustain a long-term memory recalling, sustained by neuronal circuit activated by the influx of blood from the capillary on-demand to the oxyHb release of O_2 and Mg^{2+} during transition to deoxyHb.

Mg^{2+} has been shown required for specific binding of the double-stranded DNA containing the consensus CRE sequence (5'-TGACGTCA-3').

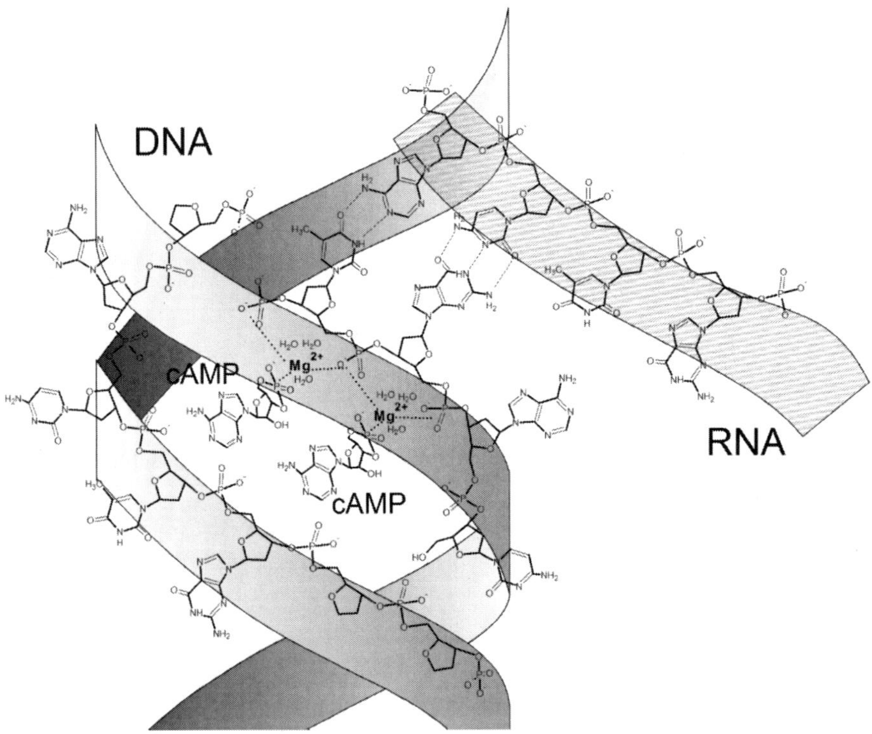

Figure 47. Crystallographic data has been used to illustrate an structure of the CREB B-ZIP protein domain, in which a hexahydrated Mg^{2+} ion binds with additional cAMP binding opens the double-stranded DNA containing the consensus CRE sequence (5'-TGACGTCA-3').

This ZIP domain could acquire the double-stranded separation because the phosphoryl group of the nucleotide of the chain would rotate to face the hexahydrated Mg^{2+}. Each one of the purine groups would be facing outwards and allow the binding of the 3'-5'-cyclicAMP through its negative charged oxygen in the cyclic configuration of its phosphoryl group. The diagram bellow illustrates that the DNA-opening has now facing to the inside each chain phosphoryl groups to bind Mg^{2+} and the purines facing to the outside to form H-bonds with the incoming nucleotide to be incorporated in one of the chains during DNA transcription [313].

The catabolite activator protein (CAP) functions by binding In the presence of the allosteric promoters and enhances the ability of RNA polymerase holoenzyme (RNAP) to bind and initiate transcription [314].

cAMP induced Mg^{2+}-dependent open DNA configuration could represent a stage in working memory the role of short-term memory intermediate.

Noradrenaline Effects on the Mind/Brain

Corpus ceruleus manufactures noradrenaline (NA) and its neurons possess long axons which branch to innervate amygdala (a center for emotions active during dreaminess), the brain stem, spinal cord, cerebellum, hypothalamus, thalamic relay nuclei and neocortex.

Social-psychic stress without adequate treatment like psychoanalysis could persist for a long-time inducing dysfunction at tissue levels like the thyroid gland.

Neural circuits are vulnerable to stress [315] sensorial neurons axons could connect with the dendrites of corpus coeruleus neurons, in which the mobilization of Ca^{2+} activates the axon release of NA vesicles within the synaptic cleft of the receiving neuron to activate the postsynaptic located adenylyl cyclase.

An overexposure of AC to NA could contribute to the dysfunctional responses characteristic of panic attacks.

The brain emotional response areas could be overstimulated, if the corpus coeruleus axons produce a large discharge of NA at the synaptic clefts without a promptly re-uptake of the neurotransmitter. Locus coeruleus (LC) neuronal losses are prominent in the pathogenesis of Alzheimer and Parkinson dysfunctions.

It is suggested that such a decrease of LC-NA, not only affects the ability to couple blood volume to oxygen demand, but also decrease the availability of NA for turning-on the activity of postsynaptic NA-stimulated adenylyl cyclase [162].

The stimulus of increased adrenaline secretion subsides only under the evolutionary advantage of psychic control, detecting the disappearance of danger. If the latter fails, the adrenaline mediated fight-or-flight response subsist. Thus, the adrenaline-stimulated increase in cortisol activates the breakdown of proteins into amino acids to support gluconeogenesis. Thus, leads into a deleterious consumption of target tissues, a characteristic of chronic metabolic stress.

The fight-or-flight tendency would be to decrease the energy reserves and flatten the circadian rhythms which synchronize ATP^{4-} availability to muscle.

The brain sympathetic system depends on NA stimulated AC and the BBB prevent adrenaline crossing into brain, a result of evolution seeking the architectural separation between body and mind/brain.

Consequently, mental stress could not be regarded as dependent *per se* of the adrenaline produced by the adrenal medulla chromaffin cells. Adrenaline by its control of glucose metabolism and its availability could affect brain functions.

The glucocorticoid released into the systemic circulation, on the other hand could cross the BBB, but a NA overstimulated hypothalamus has glucose level sensorial neurons that could be stressed by an adrenaline altered glucose metabolism. The HTPA axis could be refractive to a negative feedback by cortisol and could not turn-off the corticotropin factor (CRF) and ACTH release from the hypothalamus and pituitary respectively. Consequently, could not turn-off the adrenaline-initiated signaling of danger.

Turning-off an arousal state rest in enzymes capable to inactivate or destroy adrenaline. Other adjustments develop from the inactivation of AC by its overexposure to adrenaline.

Emotional Intelligence and Long-Term Memory Preceded the Development of Auto-Conditioned Working Memory

Many cognitive scientists support the premise that a single event could initiate learning stages differentiable by time decoding and retention, mainly short and long-term memory (LTM) [316]. The state of alertness involves sensory perceptions like vision, smell, hearing, touch, fear, starving, etc. This is a response conditionable by reward and uses the fight-or-flight widening of the brain arterioles, allowing additional nutrients like glucose, Mg^{2+} and O_2 to the region of the brain, responding with neuronal circuit activation.

The glucose sensor neurons of the hypothalamus monitor the fluctuations levels of this carbohydrate in blood. The corpus ceruleus neuronal axons discharge noradrenaline (NA) at the synapsis of receptor neurons. The long axons of the corpus coeruleus by the simultaneous discharge of NA at brain areas like cortex and corpus striatum [2, 4] could link an emotional signaling with the neurotransmitter stimulus of AC from different areas for conforming cognitive circuit [162].

It was shown that ionic Mg^{2+} networked *in common* the activatory responsiveness to NA of AC in brain and to adrenaline of AC at several body tissues. Consequently, the level of Mg^{2+} could trigger an activatory response of AC in the different tissues forming the network for simultaneous stimulation of neuronal circuits and their metabolic support [162].

A general model predicated that the trophic action of Ca^{2+} at the synaptic cleft by conjunction potentiation could support a pathway to Marr's cognitive memory [317].

Insolation of neuronal membrane and AC kinetic characterization of the purified preparation allow to describe mechanisms fitting an enzyme membrane model linking some parameters required for a system capable to process a neurotransmitter signaling into short and long-term circuit memory.

The presynaptic Ca^{2+}-dependent releases of NA by inhibiting postsynaptic AC appears as a pulsatile signaling because by replacing Ca^{2+}, the ion Mg^{2+} could activate the responsiveness to NA of postsynaptic AC.

The structural quaternary dynamics of Hb is relevant for electrogenic membrane transduction and short-term memory, because predicts that information could be coded in protein transitional states with turnover couple to coordinative chemistry and H-bond exchanges. The characterization of the responsiveness of AC may allow to postulate a modulatory mechanism for short-term memory input, into a nerve impulse. The RARE BiBi formula could be used to illustrate the oscillatory kinetic response of AC to the inhibitor ion Ca^{2+}, overcoming the activatory concentration of Mg^{2+} and viceverse.

Enzyme-Ca vs enzyme-Mg, both quaternary complexed forms of an AC, could code information by turning-off or turning-on AC-responsiveness to the hormone NA. The conjecture could describe the encoding into short-term memory as Ca^{2+} vs Mg^{2+}-dependent competition to form AC complexes. The pattern of H-bonds exchanged during interconversion between two differentiable complexes states of single polypeptide could modulate membrane vibrational states in differentiable action potentials.

The AC role in generating cyclic adenosine monophosphate (cAMP) allows a LTM through the covalent chemistry of the CREB pathway [4, 96]. Conjunction potentiation [317] results from several areas of the brain by *in common* AC-responsiveness to NA from corpus coeruleus could increase their endogenous level of cAMP to increase emotional-dependent connectivity and plasticity. The LTM role of AC had experimental support from many laboratories and has been generalized as behavior- cAMP-linked models [318, 319, 320, 321, 322].

Physiologists had explore the functional dominance of mind/brain over the body self. This may relate to the brain-localized NA-activated AC forming an enzyme-axis with the tissue networked adrenaline-activated AC of the body-self. The latter, also responds to glucagon adrenocorticotropic hormone (ACTH) for greater interdependence.

The evolution role of BBB prevents that the adrenaline in blood could reach to turn-off the signaling by the NA-neurons of brain. Thus, reserves a dominant role for the mind/brain cognition of danger disappearance. The fight-or-flight response involves suppression of the tendency to homeostasis.

A departure from instinct, the fight-or-flight response, plays the role of a choice capability conditioned by learning. This, open an evolutionary path, leading to emotional memory and intelligence. The postulation of divergent roles for nature and nurture had been blurred. A unifying perception is promoted by the finding that either stimulatory or inhibitory environmental inputs lead to changes in the strength of synaptic connectivity.

The lack of adrenaline feedback becomes self-servicing for the brain, through the fight-or-flight response to obtain emotional appropriation of the HTPA metabolic controls. This allowed the brain to expand into 10^{11} neurons, plus similar quantity of astrocytes cells, compared with the 8×10^{10} body cells. This occurs with a brain consuming calories at a rate 10 times higher than the body, per similar tissue weight. Temperature homeostasis involved to create a red of 3×10^{11} microns of capillary per neuronal-glial functional unit. Lacking a lymph system, the pulsatile astrocytes impose high turnover to the replacement of 3.7 times 160 ml/daily of cerebral spinal fluid (CSF).

Emergence of an emotional intelligence required to incorporate multiple perceptive inputs for a fight-or-flight behavior, capable to diversify the outcome response.

The evolution of the neuronal electrogenic membrane incorporates a receptor for emotional signaling to an adaptive fight-or-flight design to become the structural molecular architecture leading to cognitive memory.

Along evolution emotional memory precedes the developing of working memory and allows emotional arousal to sieve-out the parasympathetic recording and lead to the storing of experience in LTM.

Sleep is not a unitary state but the combination of two substates of distinct brain activity in different brain areas [323].

Sleep definition requires the combined input from electroencephalogram (EEG), an electrooculogram (EOG), and an electromyogram (EMG).

One state is characterized in most mammals by Rapid Eye Movements Sleep (REMS). The other stage is known as non-REM Sleep (NREMS), characterized by a virtual absence of muscle tone in antigravity muscles, a largely awake brain, and dreaming. The metabolic rate in NREMS is increased above resting waking levels.

NREMS-sleep architecture allows to recognize stages. Waking is follow by 1 and 2 NREMS's lighter stages within 15±5 minutes. Then follows the 3 and 4 Slow Wave Sleep (SWS) stages maintained for one hour in young subjects and 5-10 minutes in older ones. This cycle is repeated 3 to 4 times during the night, each decreasing SWS and increasing the REMS durations.

The first half of the night could be characterized by a surge of growth hormone (GH) secretion which takes up amino acids from blood to support muscle growth whereas ACTH and cortisol levels are maintain low.

In the course of a second half of a night, GH declines, ACTH and cortisol rise, promoting the release to blood of 1% per night of the amino acids constituting muscle, to support liver gluconeogenesis. It is suggested that heart as a muscle also releases amino acids becoming more vulnerable at this time. The early morning rise in testosterone allows re-uptake of amino acids from blood into muscle.

A ubiquitous question: why is so important the role of sleep that is present in so many phyla? Proteins became denatured by temperature. At 37°C, protein denaturation is near minimal and usually undetected. However, the breaking of proteins-H-bonds results its random reconstitution. The energy released by protein/enzymes catalytic turnover could be coupled to the breaking of H-bonds in water 1-7kcal/mol.

Heat dissipation of turnover/cycle into water occurs to recover from the intermediate states the original protein configuration. However, it could be assumed that some small percentage could become denature.

This reveals that the vital role of sleep is to replace damaged proteins and to repair brain-body affected functions.

Sleep could be characterized as a neuroendocrine regulation period with minimal processing of sensory information and lacking conspecifics stimuli from the environment. Endogenous and exogenous hormones, hormonal agents, and neuropeptides act on sleep onset or wake process. Growth hormone (GH) releasing hormone, ghrelin, galanin, neuropeptide Y promote sleep, whereas others somatostatin, corticotropin appear to impair continuity [8].

Sleep deprivation reduces learning, impairs performance in cognitive tests, prolongs reaction time, and is a common cause of seizures [324] [325].

Neuronal activity during sleep sometimes does resemble those of individuals awake. Declarative memories, newly formed, like motor skills, places, pictures, countries, etc., benefit from an afternoon's nap.

At night synaptic weakening may restore brain connections to base level to preserve synaptic homeostasis hypothesis (SHY) [326]. The night loss of alertness evidences the role of SHY. This one is to transfer during the sleeping process the ability to modify connectivity. This allows the strengthening of the recording of events, evocated by the emotional mechanism related to the noradrenergic processing of the fight-or-flight response.

Brain activity could be characterized by the day vs night responsiveness of the neuronal-astrocytes system to the Mg^{2+} vs Ca^{2+} ion signaling.

The cerebrovascular pulsation of CSF from the subarachnoid space moves rapidly into the brain astroglial water channels lining paravascular CSF pathways [327].

Influx of CSF is in part driven by arterial pulse waves that propel the movement of CSF inward along periarterial spaces [328]. Thus, arterial blood pressure is higher during physical activity. Wakefulness suppression of the influx of CSF tracers relates to an increase resistance to convective fluid movements. Hence, allowing to link the awake brain with a contraction of interstitial space, which is concordant with natural sleep states to increment by 60% interstitial space [40]. Similar results were obtained by anesthesia, obtained by intraperitoneal administration to rats of ketamine/xylazine (KX).

The clearance of potentially toxic biomolecules became strongly stimulated by sleep. An overall homeostatic control involves conservation of distance by the summing-up of the astroglial plus the interstitial space. Accordingly an increase in the interstitial space should be regarded as coupled to the shrinkage of astroglial cells [329].

Real-time application of iontophoretic tetramethyl-ammonium (TMA) to awake a head-fixed mice increased CSF influx from 14 to 23% of the interstitial volume, comparable with changes obtained during sleeping or anesthetized mice.

The electrocorticography (ECoG) method allowed to register a TMA-path dependent inhibition of adrenergic signaling is register as a change to a more sleep-like electrogenic state of brain. This finding is concordant with suppression of the responsiveness to noradrenaline could induce rapid changes in neural activity [*330*] [331]. This appears as linked to the modulation of the interstitial space [332], similar to sleep and anesthesia. This finding appears to relate the brain sleeping state, with adrenergic receptor inhibition by Ca^{2+}.

Nitrophenyl-EGTA and DM-nitrophen are Ca^{2+} cages that release Ca2+ when cleaved upon illumination with near-ultraviolet light. Photolysis releasing of caged Ca^{2+} lead to efflux of cytosolic glutamate via anion channels.

In the astrocyte the caged Ca^{2+} dependent opening of anion channels rather than exocytosis [333] triggers the glutamate release. Combined, these studies show that photolysis but not by Ca^{2+} agonists could lead to glutamate release.

Ca^{2+}-dependent release of glutamate during sleep by managing the mediating astrocyte-neuronal space could lead to modifying electroneuronal connectivity. Hence, perceived day events by 10^5 noradrenergic neurons with as much as 10^3 spines connection per LC neuron, could be transfer into the 10^{11} neurons configuring the glutaminergic system and process for selective storage.

The process may have a circadian component with specific regional responses to prevent excessive changes in the size of brain. The latter, through the process of CSF flow between the anatomical influx and efflux routes supported by astrocytic water transport. Space modification may operate like a conditioning parameter for memory transfer from the adrenergic into the glutaminergic system.

The brain emotional axis formed by LC neuron's NA, acting on the brain and the adrenal's adrenaline after secreted into blood for shifting the metabolism of somatic tissues had led to a molecular characterization. The recognizing of a molecular equivalent for the psychosomatic relationship also has social problematics impact.

These ones can be characterized by considering that the fight-or-flight response involves not only decision making, but forming a memory from the storage of the various degrees of danger and fear. The survival role of the latter, potentiated by adrenaline addiction, allows brain to sustain self-developing of functional specialized regions by its control of the soma metabolism. Thus, brain with only 2% of total tissue could divert for self-improvement, about 20% of total body energy consumption.

In terms to understand, the role of human evolution through the fight-or-flight response in terms of our specie requires that individual and group survival could be integrated through mechanisms of gratification. Accordingly to oxytocin and vasopressin favors the inclusion of the family group in survival mechanism.

The emerging of other gratification mechanisms, like endorphin, could favors social connectivity in larger groups and farther evolutionary input by group survival advantages. The finding that the brain prepares actions in anticipation of decisions has been use as an argument to challenge Human conditioning through free will.

However, this perspective of automated brain functions ignore that resulted from fight-or-flight learning implication on associated emotional memory responses. These ones allow a brain into developing mechanistic triggered response, but these are resultants from previously free will decision. Thus, the practicing of a danger sports requires a high level of auto-conditioning training. Hence, free will precedes the individual decision to reach an automatized level because the individual is seeking to experiment again the adrenaline involve in pursuing social and self-gratification.

Alphabetization results into new connections in the parietal-temporal-occipital (PTO) association areas, which link with the frontal lobe and the temporal lobe (figure 1.a). This regional connectivity does not emerge in analphabets. However, even in the adult age, by learning to read, these individuals reach the rewarding connectivity states. The increment in autoesteem allows a better fitting in society, family and results in illustrated citizens practicing of democracy and common welfare. Free will became associated to moral judgments. These processes are mediated by the prefrontal cortex, ventromedial and orbital adjacent areas (figure 1.b).

Love and desire reward circuit: Similar brain networks were activated by love and sexual desire. Love tends to annul prefrontal cortex and the sense of judgment activities, but do not decrease free will rewards.

These are mechanisms that eventually allowed the preponderance of social over instinctual behavior, which characterize our specie.

People's current social expectation from science is to receive an answer to the question: who are we? From free will decisions depend humanity survival are we going to use ecological knowledge to nurture our world?

Acknowledgments

The contributions of my PhD. Students: S. Brydon-Golz, H. Ohanian, R.H. Harris, V.A. De Bari, P.P. Vicario, and R. Cruz. K. Borhanian from special projects of the College of Medicine and Dentistry of New Jersey (CMDNJ). S. De Farias, on leave, Instituto Butantan (Sao Paulo, Brazil). N. Seidler from Rutgers under graduate projects, the asistances of E.M. De Pahn, N.A. Novak and N. Bennun and N. Ledesma, Technological University of Bs.As. are all gratefully recognized.

References

[1] Brydon-Golz, S. and Bennun, A. (1975). Postsynthetic stabilized modification of adenylate cyclase by metabolites. *Biochemical Society Transactions*, 3, 721-724.

[2] Brydon-Golz, S.; Ohanian, H. and Bennun, A. (1977). Effects of noradrenaline on the activation and the stability of brain adenylate cyclase. *Biochem. J.*, 166, 473-483.

[3] Lowry, O. H.; Rosebrough, N. J.; Farr, A. L. and Randall R. J. (1951). Protein measurement with the Folin phenol reagent. *J. Biol. Chem.*, 193, 265-275.

[4] Ohanian, H.; Borhanian, K.; De Farias, S. and Bennun, A. (1981) A model for the regulation of brain adenylate cyclase by ionic equilibria. *Journal of Bioenergetics and Biomembranes*, Vol. 13, 5/6, 317-55.

[5] Gilman, A. G. (1970). A protein binding assay for adenosine 3,5-cyclic monophosphate. *Proc. Natl. Acad. Sci.*, 67(1), 305-312.

[6] Harris, R.; Cruz, R. and Bennun, A. (1979). The effect of hormones on metal and metal-ATP interactions with fat cell adenylate cyclase. *BioSystems*, 11, 29-46.

[7] Rodbell, M. (1964) Metabolism of isolated fat cells. I. effects of hormones on glucose metabolism and lipolysis. *J. Biol. Chem.*, 239, 375-80.

[8] Harris, R. and Bennun, A. (1976). Hormonal control of fat cells adenylate cyclase. *Molecular & Cellular Biochemistry*, 13 (3), 141-146.

[9] Brown, B. L.; Ekins, R. P. and Albano, J. D. M. (1972). Saturation assay for cyclic AMP using endogenous binding protein. *In Advances in Cyclic Nucleotide Research*, 2, 25-40.

[10] Bennun, A. and Racker, E. (1969). Partial resolution of the enzymes catalyzing photophospharylation IV. Interaction of coupling factor I from chloroplast with components of the chloroplast membrane. *J. Biol. Chem.*, 244, 1325-1331.

[11] Miyamoto, E.; Kuo, J. F. and Greengard, P. J. (1969). Cyclic nucleotide-dependent protein kinases. 3. Purification and properties of adenosine 3',5'-monophosphate-dependent protein kinase from bovine brain. *J. Biol. Chem.*, 244 (23), 6395-6402.

[12] Bray, G. A. (1960). A simple efficient liquid scintillator for counting aqueous solutions in a liquid scintillation counter. *Anal. Biochem.* 1, 279-285.

[13] Klein, H. H.; Friedenberg G. R.; Cordera R. and Olefsky J. (1985). Substrate specificities of insulin and epidermal growth factor receptor kinases. *Biochem. Biophys. Res. Commun.*, 127, 254-263.

[14] Vicario, P. P.; Saperstein R. and Bennun, A. (1988). Role of divalent metals in the activation and regulation of insulin receptor tyrosine kinase. Biosystems., 22(1), 55-66.

[15] Pike, L. J.; Kuenzel, E. A.; Casnelli, J. E. and Krebs, E. G. (1984). A comparison of the insulin- and epidermal growth factor-stimulated protein kinases from human placenta. *J. Biol. Chem.*, 259, 9913-9921.

[16] Cleland, W. W. (1963). The kinetics of enzyme-catalyzed reactions with two or more substrates or products. I. Nomenclature and rate equations. *Biochim. Biophys., Acta*, 67, 104-37, 173-87.

[17] Teo, I.; Yeow Chin, K; Stephens, C. and Paget J. (2014). Drugs in Cardiopulmonary Resuscitation. *Book of Medicine, Endocrine Disorders. Editorial Nova Science Publishers. Editor: Alfred Bennun. Adrenaline: Production, Role in Disease and Stress, Effects on the Mind and Body*, pp. 177-212.

[18] Chowdhury, T.; Sandu, N. and Schaller, B. (2014). New Insights to the Role of (Nor-)/Adrenaline and Trigeminal Cardiac Reflex. *Book of Medicine, Endocrine Disorders. Editorial Nova Science Publishers. Editor: Alfred Bennun. Adrenaline: Production, Role in Disease and Stress, Effects on the Mind and Body,* pp. 71-80.

[19] Baker, C. N.; Katsandris, R.; Van, C. and Ebert, S. N. (2014). Adrenaline and Stress-Induced Cardiomyopathies: Three Competing Hypotheses for Mechanism(s) of Action. *Book of Medicine, Endocrine Disorders. Editorial Nova Science Publishers. Editor: Alfred Bennun. Adrenaline: Production, Role in Disease and Stress, Effects on the Mind and Body*, pp. 81-116.

[20] Myslivecek, J.; Valuskova, P. and Varejkova, E. (2014). Adrenaline, Heart Adrenoceptors and Stress. Endocrine Disorders. *Book of Medicine, Endocrine Disorders. Editorial Nova Science Publishers. Editor: Alfred Bennun. Adrenaline: Production, Role in Disease and Stress, Effects on the Mind and Body*, pp. 117-148.

[21] Baba, Y.; Hayashi, S.; Ikeda, S. and Nakajo, M. (2014). Adrenalines in adrenal venous sampling. *Book of Medicine, Endocrine Disorders. Editorial Nova Science Publishers. Editor: Alfred Bennun. Adrenaline: Production, Role in Disease and Stress, Effects on the Mind and Body*, pp. 1-12, 2014.

[22] Brou, C.; Logeat, F.; Gupta, N; Bessia C.; LeBail, O.; Doedens, J. R.; Cumano, A.; Roux, P.; Black, R. A. and Israël, A. (2000). A novel proteolytic cleavage involved in notch signaling: the role of the disintegrin-metalloprotease TACE. *Mol. Cell*, 5 (2), 207-16.

[23] Oliva, C. A.; Vargas, J. Y. and Inestrosa, N. C. (2013). Wnts in adult brain: from synaptic plasticity to cognitive deficiencies. *Front Cell Neurosci*. 7, 224.

[24] Dehay, C.; Douglas, R. J.; Martin, K. A. and Nelson, C. (1991). Excitation by geniculocortical synapses is not 'vetoed' at the level of dendritic spines in cat visual cortex. *J. Physiol.*, 440, 723-34.

[25] Knott, G. W.; Quairiaux, C.; Genoud, C. and Welker, E. (2002). Formation of dendritic spines with GABAergic synapses induced by whisker stimulation in adult mice. *Neuron.*, 34 (2), 265-273.

[26] Wilson, C. J.; Croves, P. M.; Kitai, S. T. and Linder, J. C. (1983). Three-dimensional structure of dendritic spines in the rat neostriatum. *J. Neurosci.*, 3, 383-388.

[27] Crivellano, E.; Nico, B. and Ribatti, D. (2005) Ultrastructural evidence of piecemeal degranulation in large dense-core vesicles of brain neurons. *Anat. Embryol. (Berl)*, 210 (1), 25-34.

[28] Torrealba, F. and Carrasco, M. A. (2004). A review on electron microscopy and neurotransmitter systems. *Brain Res. Brain Res. Rev.*, 47 (1-3), 5-17.

[29] Ahmari, S. E. and Smith, S. J. (2002). Knowing a nascent synapse when you see it. *Neuron*, 34 (3), 333-336.

[30] Dahlstroem, A. and Fuxe, K. (1964). Evidence for the existence of monoamine-containing neurons in the central nervous system. I. demonstration of monoamines in the cell bodies of brain stem neurons. *Acta Physiol Scand Suppl.*, 232, 1-55.

[31] Ward, D. G. and Gunn, C. G. (1976). Locus coeruleus complex: elicitation of a pressor response and a brain stem region necessary for its occurrence. *Brain Res.* 107 (2), 401-406.

[32] Halliday, G. (2004). Substantia nigra and locus Coeruleus. *In: Paxinos G, Mai JK, editors. The Human Nervous System - II Edition. San Diego (CA): Elsevier Academic Press*; pp. 449-463.

[33] Aston-Jones G. S.; Iba, M.; Clayton, E.; Rajkowski, J. and Cohen, J. (2007). Locus coeruleus and regulation of behavioral flexibility and attention: clinical implications. *In: Ordway GA, Schwartz MA, Frazer A, editors. Brain Norepinephrine - Neurobiology and Therapeutics. Cambridge: University Press*, pp. 196-235.

[34] Iversen, L. L.; Rossor, M. N.; Reynolds, G. P.; Hills, R.; Roth, M.; Mountjoy, C. Q.; Foote, S. L.; Morrison, J. H. and Bloom, F. E. (1983). Loss of pigmented dopamine-beta-hydroxylase positive cells from locus coeruleus in senile dementia of Alzheimer's type. *Neurosci Lett.*, 39 (1), 95-100.

[35] Baker, K. G.; Törk, I; Hornung, J. P. and Halasz P. (1989). The human locus coeruleus complex: an immunohistochemical and three dimensional reconstruction study. *Exp Brain Res.*, 77 (2), 257-270.

[36] Wright, E. E. and Simpson, E. R. (1981). Inhibition of the lipolytic action of beta-adrenergic agonists in human adipocytes by alpha-adrenergic agonists. *J. Lipid Res.*, 22 (8), 1265-1270.

[37] Samuels, E. R. and Szabadi, E. (2008). Functional Neuroanatomy of the Noradrenergic Locus Coeruleus: Its Roles in the Regulation of Arousal and Autonomic Function Part I: Principles of Functional Organisation. *Curr Neuropharmacol.*, 6(3), 235-253.

[38] Mouton, P. R.; Pakkenberg, B.; Gundersen, H. J.; Price, D. L. (1994). Absolute number and size of pigmented locus coeruleus neurons in young and aged individuals. *J. Chem. Neuroanat.*, 7 (3), 185-190.

[39] Kety, S. S. (1970). The biogenic amines in the central nervous system: their possible roles in arousal, emotion and learning. *In the neurosciences: second study program*, 324-335.

[40] Xie, L.; Kang, H.; Xu, Q.; Chen, M. J.; Liao, Y.; Thiyagarajan, M.; O'Donnell, J.; Christensen, D. J.; Nicholson, C.; Iliff, J. J.; Takano, T.; Deane, R. and Nedergaard M. (2013). Sleep drives metabolite clearance from the adult brain. *Science.*, 342 (6156), 373-377.

[41] Silvetti, M.; Seurinck, R.; van Bochove, M. E. and Verguts, T. (2013). The influence of the noradrenergic system on optimal control of neural plasticity. *Front Behav Neurosci.*, 7, 160.

[42] Shankar, E.; Santhosh, K. T. and Paulose, C. S. (2006). Dopaminergic regulation of glucose-induced insulin secretion through dopamine D2 receptors in the pancreatic islets in vitro. *IUBMB Life*, 58(3), 157-63.

[43] Rubí, B.; Ljubicic, S.; Pournourmohammadi, S.; Carobbio, S.; Armanet, M.; Bartley, C. and Maechler, P. (2005). Dopamine D2-like receptors are expressed in pancreatic beta cells and mediate inhibition of insulin secretion. *J. Biol. Chem.*, 280(44), 36824-32.

[44] Humeau, Y.; Doussau, F.; Grant, N. J. and Poulain, B. (2000). How botulinum and tetanus neurotoxins block neurotransmitter release. *Biochimie.*, 82 (5), 427-446.

[45] Kaehler, S. T.; Singewald, N. and Philippu, A. (1999). Dependence of serotonin release in the locus coeruleus on dorsal raphe neuronal activity. *Naunyn Schmiedebergs Arch Pharmacol.*, 359 (5), 386-93.

[46] Colonnier, M. (1968). Synaptic patterns on different cell types in the different laminae of the cat visual cortex. An electron microscope study. *Brain Res.*, 9, 268-287.

[47] Cooney, J. R.; Hurlburt, J. L.; Selig, D. K., Harris, K. M. and Fiala, J. C. (2002) Endosomal compartments serve multiple hippocampal dendritic spines from a widespreas rather than a local store of recycling membrane. *J. Neurosci.*, 22, 2215-24.

[48] Cornell-Bell, A. H.; Thomas, P. G. and Smith, S. J. (1990). The excitatory neurotransmitter glutamate causes filopodia formation in cultured hippocampal astrocytes. *Glia*, 3, 322-334.

[49] Berger, M.; Gray, J. A and Roth, B. L. (2009). The expanded biology of serotonin. *Annu. Rev. Med.*, 60, 355–66.

[50] Young, S. N. (2007). How to increase serotonin in the human brain without drugs. *Rev. Psychiatr. Neurosci.*, 32 (6), 394-99.

[51] Carlson, N. R. (2012). Physiology of behavior. Pearson, pp. 336.

[52] Solano-Castiella, E.; Anwander, A.; Lohmann, G.; Weiss, M.; Docherty, C.; Geyer, S.; Reimer, E.; Friederici, A. D. and Turner, R. (2010). Diffusion tensor imaging segments the human amygdala in vivo. *Neuroimage*, 49 (4), 2958–65.

[53] McDannald, M.; Kerfoot, E.; Gallagher, M. and Holland, P. C. (2005). Amygdala central nucleus function is necessary for learning, but not expression of conditioned auditory orienting. *Behavioral Neuroscience*, 119 (1), 202-212.

[54] Blumberg, H. P.; Kaufman, J; Martin, A; Whiteman, R; Zhang, J. H.; Gore, J. C.; Charney, D. S. ; Krystal, J. H. and Peterson, B. S. (2003).

Amygdala and hippocampal volumes in adolescents and adults with bipolar disorder. *Arch Gen Psychiatry*, 60 (12), 1201–8.

[55] Blair, R. J. R. (2008). The amygdala and ventromedial prefrontal cortex: functional contributions and dysfunction in psychopathy. *Philosophical Transactions of the Royal Society,* 363 (1503), 2557–2565.

[56] Sheline Y. I.; Barch, D. M.; Donnelly, J. M.; Ollinger, J. M.; Snyder, A. Z. and Mintun, M. A. (2001). Increased amygdala response to masked emotional faces in depressed subjects resolves with antidepressant treatment: an fMRI study. *Biological Psychiatry,* 50 (9), 651-658.

[57] Martin, A. and Chao, L. L. (2001). Semantic memory and the brain: structures and processes. *Current Opinion in Neurobiology*, 11 (2), 194-201.

[58] Szepietowska, B.; Zhu, W.; Chan, O.; Horblitt, A.; Dziura, J. and Sherwin, R. S. (2011). Modulation of β-Adrenergic Receptors in the Ventromedial Hypothalamus Influences Counterregulatory Responses to Hypoglycemia. *Diabetes*., 60 (12), 3154-8.

[59] Bookout, A. L.; de Groot, M. H. M.; Owen, B. M.; Lee, S.; Gautron, L.; Lawrence, H. L.; Ding, X.; Elmquist, J. K.; Takahashi, J. S.; Mangelsdorf, D. J. and Kliewer, S. A. (2013). FGF21 regulates circadian behavior and metabolism by acting on the nervous system. *Nat Med.*, 19 (9), 1147-1152.

[60] Babar, S. M. (2013). SIADH associated with ciprofloxacin. *Ann Pharmacother*, 47 (10), 1359-63.

[61] Caldwell, H. K., and Young W. S. III (2006). Oxytocin and Vasopressin: Genetics and Behavioral Implications. In Lajtha A, Lim R. Handbook of Neurochemistry and Molecular Neurobiology: Neuroactive Proteins and Peptides (3rd ed.). Berlin: Springer. pp. 573–607.

[62] Nielsen, S.; Chou, C. L.; Marples, D.; Christensen, E. I.; Kishore, B. K. and Knepper, M. A. (1995). Vasopressin increases water permeability of kidney collecting duct by inducing translocation of aquaporin-CD water channels to plasma membrane. *Proc. Natl. Acad. Sci. U.S.A.*, 92 (4), 1013-7.

[63] Marieb, E. N. and Hoehn K. N. (2012). Human Anatomy & Physiology 9th edition, chapter:16, page:599. Series: Books a la Carte. Publisher: Pearson.

[64] Lee, H. J.; Macbeth, A. H.; Pagani, J. H. and Young, W. S. (2009). Oxytocin: the great facilitator of life. *Prog. Neurobiol.*, 88 (2), 127-51.

[65] De Dreu, C. K.; Greer, L. L.; Van Kleef G. A.; Shalvi, S. and Handgraaf, M. J. (2011). Oxytocin promotes human ethnocentrism. *Proc. Natl. Acad. Sci. U.S.A.*, 108 (4), 1262-6.

[66] Malik, A. I.; Zai, C. C.; Abu, Z.; Nowrouzi, B. and Beitchman, J. H. (2012). The role of oxytocin and oxytocin receptor gene variants in childhood-onset aggression. *Genes Brain Behav.*, 11 (5), 545–51.

[67] Kalat, J. W. (2001). Biological psychology (7th ed.). Belmont, CA: Wadsworth Publishing.

[68] Krupic, J.; Burgess, N. and O'Keefe J. (2012). Neural representations of location composed of spatially periodic bands. *Science*. 17, 337(6096), 853-7.

[69] Krupic, J.; Bauza, M.; Burton, S.; Lever, C. and O'Keefe, J. (2013). How environment geometry affects grid cell symmetry and what we can learn from it. *Philos Trans R Soc Lond B Biol Sci.*, 23, 369(1635), 20130188.

[70] Bukalo, O.; Campanac, E.; Hoffman, D. A. and Fields, R. D. (2013). Synaptic plasticity by antidromic firing during hippocampal network oscillations. *Proc Natl Acad Sci U.S.A.*, 110(13), 5175-80.

[71] Sadeh, T.; Shohamy, D.; Levy, D.R.; Reggev, N. and Maril, A. (2010). Cooperation between the hippocampus and the striatum during episodic encoding. *J. Cogn. Neurosci.*, 23 (7), 1597-608.

[72] Tulving, E. and Markowitsch, H. J. (1998). Episodic and declarative memory: role of the hippocampus. *Hippocampus*, 8 (3), 198-204.

[73] Ziv, N. E.; Garner, C. C. (2004). Cellular and molecular mechanisms of presynaptic assembly. *Nat. Rev. Neurosci.*, 5, 385-99.

[74] Sorra, K. E.; Mishra, A.; Kirov, S. A. and Harris K. M. (2006) Dense core vesicles resemble active-zone transport vesicles and are dimished following synaptogenesis in mature hippocampal slices. *Neuroscience*, 141, 2097-106.

[75] Shepherd, G. M. and Harris K.M. (1998). Three-dimensional structure and composition of CA3--> CA1 axons in rat hippocampal slices: implications for presynaptic connectivity and compartmentalization. *J. Neurosci.* 18, 8300-10.

[76] Sorra, K. E. and Harris, K. M. (1993). Occurrence and three-dimensional structure of multiple synapses between individual radiatum axons and their target pyramidal cells in hippocampal area CA1. *J. Neurosci.*, 13, 3736-48.

[77] Yankova, M. and Hart, S. A. (2001). Woolley C.S. Estrogen increases synaptic connectivity between single presynaptic inputs and multiple

postsynaptic CA1 pyramidal cells: a serial electron-microscopic study. *Proc. Natl. Acad. Sci. USA*, 98, 3525-30.

[78] Corkin, S.; Amaral, D. G.; Gonzalez, R. G.; Johnson, K. A. and Hyman, B. T. (1997). H.M.'s medial temporal lobe lesion: Findings from magnetic resonance imaging. *The Journal of Neuroscience*, 17, 3964-3979.

[79] Zola-Morgan, S.; Suire, L. R. (1993). Neuroanatomy of memory. *Annual Reviews Neuroscience*, 16, 547-563.

[80] Mertor, R. (1996). Memory of time may be factor in Parkinson's. Columbia University Record, Vol 21, No. 22.

[81] Forgetfulness is the Key to a Healthy Mind. *New Scientist*, February 16. 2008.

[82] Bennun, A. (1974) The unitary hypothesis on the coupling of energy transduction and its relevance to the modeling of mechanisms. *Annals of the New York Academy of Sciences*, 227, 116-145.

[83] Bennun, A. (1971). Hypothesis for coupling energy transduction with ATP synthesis or ATP hydrolysis. *Nature New Biology*, 233 (35), 5-8.

[84] Bennun, A. (1975). Hypothesis on the role of liganded states of proteins in energy transducing systems. *BioSytems*, 7, 230-244.

[85] Cahill, L. and McGaugh, J. L. (1998). Mechanisms of emotional arousal and lasting declarative memory. *Trends Neurosci.*, 21, 294-9.

[86] Costa-Mattioli, M. and Sonenberg, N. (2008). Translational control of gene expression: a molecular switch for memory storage. *Prog. Brain Res.*, 169, 81-95.

[87] Grau, C.; Ginhoux, R.; Riera, A.; Nguyen, T. L.; Chauvat, H.; Berg, M.; Amengual, J. L.; Pascual-Leone, A. and Ruffini, G. (2014). Conscious Brain-to-Brain Communication in Humans Using Non-Invasive Technologies. DOI: 10.1371/journal.pone.0105225.

[88] Amunts, K.; Kedo, O.; Kindler, M.; Pieperhoff, P.; Mohlberg, H.; Shah, N.; Habel, U.; Schneider, F. and Zilles, K. (2005). Cytoarchitectonic mapping of the human amygdala, hippocampal region and entorhinal cortex: intersubject variability and probability maps. *Anat Embryol (Berl)*, 210 (5-6), 343-52.

[89] Carlson, N. R. (1/12/2012). *Physiology of Behavior. Pearson*. p. 364.

[90] Petrovich, G. D.; Ross, C. A.; Mody, P.; Holland, P. C. and Gallagherw, M. (2009). Central, but not basolateral amygdala, is critical for control of feeding by aversive learned cues. *J Neurosci.*, 29(48), 15205–15212.

[91] Hutcherson, C. A.; Seppala, E. M. and Gross, J. J. (2008). Loving-kindness meditation increases social connectedness. *Emotion*, 8 (5), 720-4.

[92] Paré, D.; Collins, D. R. and Pelletier, J. G. (2002). Amygdala oscillations and the consolidation of emotional memories. *Trends in Cognitive Sciences*, 6 (7), 306-14.

[93] Asari, T.; Konishi, S.; Jimura, K.; Chikazoe, J.; Nakamura, N. and Miyashita, Y. (2010). Amygdalar enlargement associated with unique perception. *Cortex*, 46 (1), 94-99.

[94] Maren, S. (1999). Long-term potentiation in the amygdala: a mechanism for emotional learning and memory. *Trends Neurosci*, 22 (12): 561–7.

[95] Bennun, A.; Seidler, N. and De Bari, V. A. (1985). A model for the regulation of haemoglobin affinity for oxygen. *Biochemical Society Transactions*, 13, 364-366.

[96] Bennun, A. (2010). Characterization of the norepinephrine-activation of adenylate cyclase suggests a role in memory affirmation pathways. Overexposure to epinephrine inactivates adenylate cyclase, a causal pathway for stress-pathologies. *Biosystems*, 100(2), 87-93.

[97] Mure, H.; Hirano, S.; Tang, C. C.; Isaias, I. U.; Antonini, A.; Ma, Y.; Dhawan, V. and Eidelberg, D. (2011). Parkinson's Disease Tremor-Related Metabolic Network: Characterization, Progression, and Treatment Effects. *Neuroimage*., 54(2), 1244-53.

[98] Pavlov, M.; Siegbahn, P. and Sandström, M. (1998). Hydration of Beryllium, Magnesium, calcium, and zinc ions using density functional theory. *J. Phys. Chem. A.*, 102 (1), 219-28.

[99] Bennun, A.; Seidler, N. and De Bari, V. A. (1986). Divalent metals in the regulation of hemoglobin affinity for oxygen. *Ann. N. Y. Acad. Sci.*, 463, 76-79.

[100] Bennun, A.; Needle, N. A. and De Bari, V. A. (1985). Infrared spectroscopy of erythrocyte plasma membranes. *Biochem. Soc. Trans.*, 13, 127-128.

[101] Stoppani, A.O.M.; Bennun, A. and De Pahn, E.M. (1964). Energy requirement for the anaerobic oxidation of acetate in baker's yeast. *Biochimica et Biophysica Acta*, 92, 176-178.

[102] Bennun, A.; De Pahn, E. M. and Stoppani, A.O.M. (1964). Some properties of particle-bound intracellular ATPase from baker's yeast. *Biochimica et Biophysica Acta*, 89, 532-539.

[103] Stoppani, A.O.M.; Bennun, A. and De Pahn, E.M. (1964). Effect of 2,4-dinitrophenol on krebs cycle and phosphate metabolism in baker's yeast. *Archives of Biochemistry and Biophysics*, 108(2), 258-265.

[104] Yamazaki, S.; Numano, R.; Abe, M.; Hida, A.; Takahashi, R.; Ueda, M.; Block, G. D.; Sakaki, Y.; Menaker, M. and Tei, H. (2000). Resetting central and peripheral circadian oscillators in transgenic rats. *Science.*, 288(5466), 682-5.

[105] Chung, S.; Lee, E. J.; Yun, S.; Choe, H. K.; Park, S. B.; Son, H. J.; Kim, K. S.; Dluzen, D. E.; Lee, I.; Hwang, O.; Son, G. H. and Kim, K. (2014). Impact of circadian nuclear receptor REV-ERBα on midbrain dopamine production and mood regulation. *Cell*, 157(4), 858-68.

[106] Schibler, U. and Sassone-Corsi P. (2002) A web of circadian pacemakers. *Cell*, 111 (7), 919-22.

[107] Toh, K. L.; Jones, C. R.; He, Y.; Eide, E. J.; Hinz, W. A.; Virshup, D. M.; Ptácek, L. J. and Fu, Y. H. (2001). An hPer2 phosphorylation site mutation in familial advanced sleep phase syndrome. *Science*. 291(5506), 1040-3.

[108] Taheri, S. and Mignot, E. (2002). The genetics of sleep disorders. *Lancet Neurol.*, 1 (4), 242-50.

[109] Martin, S. R.; Masino, L. and Bayley, P. M. (2000) Enhancement by Mg^{2+} of domain specificity in Ca^{2+}-dependent interactions of calmodulin with target sequences. *Protein Sci.*, 9(12), 2477-88.

[110] Kleene, R.; Mzoughi, M.; Joshi, G.; Kalus, I.; Bormann, U.; Schulze, C.; Xiao, M. F.; Dityatev, A. and Schachner, M. (2010). NCAM-induced neurite outgrowth depends on binding of calmodulin to NCAM and on nuclear import of NCAM and fak fragments. *J. Neurosci.*, 30(32), 10784-98.

[111] Jurado, L. A.; Chockalingam, P. S. and Jarrett, H. W. (1999). Apocalmodulin. *Physiol. Rev*. 79(3), 661-82.

[112] Malmendal, A.; Linse, S.; Evenäs, J.; Forsén, S. and Drakenberg, T. (1999). Battle for the EF-hands: magnesium-calcium interference in calmodulin. *Biochemistry.*, 38(36), 11844-50.

[113] Guther, T. and Vormann, J. (1992). Na^{+}-dependent Mg^{2+} efflux from isolated perfused rat hearts. *Magnes Bull.*, 14, 126-129.

[114] Guther, T.; Vormann, J. and Hollriegl, V. (1991) Noradrenaline-induced Na^{+}-dependent Mg^{2+} efflux from rat liver. *Magnes Bull.*, 13, 122-124.

[115] Romani, A.; Marfella, C. and Scarpa, A. (1993). Regulation of magnesium uptake and release in the heart and in isolated ventricular myocytes. *Circ Res.*, 72, 1139-1148.

[116] Takano, T. and Nedergaard, M. (2007). DM-nitrophen provides physiologically significant increases of intracellular [Ca^{2+}] in the presence of Mg^{2+}. *Cell Calcium.*, 41 (5), 503-4.

[117] Romani, A. M. P.; Matthews, V. D. and Scarpa, A. (2000). Parallel stimulation of glucose and Mg2+ accumulation by insulin in rat hearts and cardiac ventricular myocytes. *Circ Res.*, 86, 326-333.

[118] Romani, A. and Scarpa, A. (1990). Hormonal control of Mg^{2+} transport in the heart. *Nature*, 346, 841-844.

[119] Flatman P. W. (1980). The effect of buffer composition and deoxygenation on the concentration of ionized magnesium inside human red blood cells. *J. Physiol.*, 300, 19-30.

[120] Bunn, H. F.; Ransil, B. J. and Chao, A. (1971). The interaction between erythrocyte organic phosphates, magnesium ion and hemoglobin. *J. Biol. Chem.* 246, 5273-5279.

[121] Somlyo, A. V.; McClellan, G.; Gonzalaz-Serratos, H. and Somlyo, A. P. (1985). Electron probe X-ray microanalysis of post-tetanic Ca^{2+} and Mg^{2+} movements across the sarcoplasmic reticulum in situ. *J. Biol. Chem.*, 260, 6801-6807.

[122] Munshi, P.; Stanley, C. B.; Ghimire-Rijal, S.; Lu, X.; Myles, D. A. and Cuneo, M. J. (2013). Molecular details of ligand selectivity determinants in a promiscuous β-glucan periplasmic binding protein. *BMC Struct Biol.*, 13, 18.

[123] Skou, J. C. (1957). The influence of some cations on an adenosine triphosphatase from peripheral nerves. *Biochim. Biophys. Acta*, 23, 394 (1957).

[124] Garrahan P. J. and Glynn I. M. (1966). Driving the sodium pump backwards to form adenosine triphosphate. *Nature*, 211 (5056), 1414-5.

[125] Bennun, A. (1971) Interaction of the chloroplast coupling factor with protons and water, Congreso Argentino de Ciencias Biológicas - 1970, in "Recientes adelantos en Biología" (J.A. Moguilevsky and R. Mejía, eds.), pp. 254-264, University of Buenos Aires Press.

[126] Bennun, A. (1971). Properties of chloroplast's coupling factor-1 and a hypothesis for a mechanism of energy transduction, Proceedings First European Biophysics Congress, Baden, Austria, 1971, in "Photosynthesis, bioenergetics, regulation, origin of life" (E. Broda, A. Locker and H. Sprínger-Lederer, eds.), IV, 85-91, Wiener Medizinischen Akademíe, Vienna.

[127] Bennun, A. (2013). The coupling of thermodynamics with the organizational water-protein intra-dynamics driven by the H-bonds dissipative potential of cluster water. *arXiv: 1303.6993 [q-bio.MN].*

[128] Belton, P.S. (2000). Nuclear magnetic resonance studies of the hydration of proteins and DNA. *Cell. Mol. Life Sci.*, 57, 993–998.

[129] Kornblatt, J. and Kornblatt, J. (1997). The role of water in recognition and catalysis by enzymes. *The Biochemist*, 19 (3), 14–17.

[130] Servantie, J.; Atilgan, C. and Atilgan A. R. (2010). Depth dependent dynamics in the hydration shell of a protein. *J. Chem Phys.*, 133 (8), 085101.

[131] Tournier, A. L.; Xu, J. and Smith, J. C. (2003). Translational hydration water dynamics drives the protein glass transition. *Biophys. J.*, 85 (3), 1871-5.

[132] Chanutin, A. and Curnish, R. R. (1967). Effect of organic and inorganic phosphates on the oxygen equilibrium of human erythrocytes. *Arch. Biochem. Biophys.*, 121(1), 96-102.

[133] Arnone, A. (1972). X-ray diffraction study of binding of 2,3-diphosphoglycerate to human deoxyhaemoglobin. *Nature*, 237, 146-149.

[134] Antonini, E.and Brunori, M. (1969). On the rate of a conformation change associated with ligand binding in hemoglobin. *J. Biol. Chem.*, 244 (14), 3909-3912.

[135] Benesch, R. E. (1994). The stability of the heme-globin linkage: measurement of heme exchange. *Methods Enzymol.*, 231, 496-502.

[136] Benesch, R. and Benesch, R. E. (1967). The effect of organic phosphates from the human erythrocyte on the allosteric properties of hemoglobin. *Biochem. Biophys. Res. Commun.*, 26, 162-174.

[137] Gerber, G.; Berger, H.; Janig, G. R. and Rapoport, S. M. (1973) Interaction of haemoglobin with ions. Quantitative description of the state of magnesium, adenosine 5'-triphosphate, 2,3-bisphosphoglycerate, and human haemoglobin under simulated intracellular conditions. *Eur. J. Biochem.*, 38, 563-571.

[138] Vicario, P. P. and Bennun, A. (1990). Separate effects of Mg2+, MgATP, and ATP4- on the kinetic mechanism for insulin receptor tyrosine kinase. *Archives of Biochemistry and Biophysics*, 278, 99-105.

[139] Fermi, G. and Perutz, M. F. (1982) In: Atlas of molecular structures in Biology, 2 Haemoglobin and Myoglobin D.C., Phillips and Richards F.M. (Eds.) Clarendon Press, Oxford.

[140] Rifkind, J. M. and Heim, J. M. (1977). Interaction of zinc and hemoglobin: binding of zinc and the oxygen affinity. *Biochemistry*, 16, 4438-4443.

[141] Benesch, R. and Benesch R. E. (1969). Intracellular organic phosphates as regulators of oxygen release by hemoglobin. *Nature*, 221, 618-622.

[142] Perutz, M. F. (1970). Stereochemistry of cooperative effects in heamoglobin. *Nature*, 228, 726-739.

[143] Perutz, M. F.; Wilkinson, A. J.; Paoli, M. and Dodson, G. G. (1998). The stereochemical mechanism of the cooperative effects in hemoglobin revisited. *Annu. Rev. Biophys. Biomol. Struct.*, 27, 1-34.

[144] Panin L. E. (2014). Influence of stress hormones (adrenaline and cortisol) on structure and function of erythrocyte membranes. *Book of Medicine, Endocrine Disorders. Editorial Nova Science Publishers. Editor: Alfred Bennun. Adrenaline: Production, Role in Disease and Stress, Effects on the Mind and Body*, pp. 149-176.

[145] Churchill, P. C. and Churchill, M. C. (1982). Ca-dependence of the inhibitory effect of K-depolarization on renin secretion from rat kidney slices. *Arch. Int. Pharmacodyn. Ther.*, 258 (2), 300-312.

[146] Perutz, M. F.; Paoli, M. and Lesk, A. M. (1999). Fix L, a haemoglobin that acts as an oxygen sensor: signalling mechanism and structural basis of its homology with PAS domains. *Chem. Biol.*, 6, 291-7.

[147] Ebbinghaus, S.; Kim, S. J.; Heyden, M.; Yu, X.; Heugen, U.; Gruebele, M.; Leitner, D. M. and Havenith, M. (2007). An extended dynamical hydration shell around proteins. *Proc. Natl. Acad. Sci. USA*. 104 (52), 20749-20752.

[148] Bennun, A. (1987). A coupling mechanism to inter-relate regulatory with haem-haem interactions of haemoglobin. *Biomed. Biochim. Acta*, 46 (2-3), S314-9.

[149] Klug, A. and Rhodes, D. (1987). Zinc fingers: a novel protein fold for nucleic acid recognition. *Cold Spring Harb. Symp. Quant. Biol.*, 52, 473-82.

[150] McGrath, C. F.; Buckman, J. S.; Gagliardi, T. D.; Bosche, W. J.; Coren, L. V. and Gorelick, R. J. (2003). Human cellular nucleic acid-binding protein Zn^{2+} fingers support replication of human immunodeficiency virus type 1 when they are substituted in the nucleocapsid protein. *J. Virol.*, 77(15), 8524-31.

[151] Deuticke, B.; Duhm, J. and Dierkesmann, R. (1971). Maximal elevation of 2,3-diphosphoglycerate concentrations in human erythrocytes:

Influence on glycolytic metabolism and intracellular Ph. *Pflugers Arch.*, 326 (1), 15-34.

[152] Brewer, G. J. (1974). General red cell metabolism. *In: Surgenor D. ed. The red blood cell. New York: Academic Press*, 387-433.

[153] Thomas, E. L.; King, L. E. and Morrison, M. (1979). The uptake of cyclic AMP by human erythrocytes and its effect of membrane phosphorylation. *Archs. Biochem. Biophys.*, 196, 459-464.

[154] DeBari, V. A. and Bennun, A. (1982). Cyclic GMP in the human erythrocyte. Intracellular levels and transport in normal subjects and chronic hemodialysis patients. *Clin. Biochem.*, 15, 219-221.

[155] DeBari, V. A.; Novak, N. A. and Bennun, A. (1984). Cyclic Nucleotide Metabolism in the Human Erythrocyte. *Clin. Physiol. Biochem.*, 2, 227-238.

[156] Novembre, P.; Nicotra, J.; DeBari, V. A.; Needle, M. A. and Bennun, A. (1984). Erythrocyte transport of cyclic nucleotides. *Annals of the New York Academy of Sciences*, 435, 190-194.

[157] Marie, J.; Tichonicky, L.; Dreyfus, J. C. and Kahn, A. (1979). Endogenous, cyclic 3',5' AMP-dependent phosphorylation of Human red cell pyruvate kinase. *Bichem. Biophys. Res. Commun.*, 87, 862-867.

[158] Vicario P. P. and Bennun, A. (1989). Interaction of MnATP and peptide substrate with insulin receptor tyrosine kinase. *Biochem. Soc. Trans.*, 17 (6), 1108-9.

[159] Vicario P. P. and Bennun, A. (1989). Regulation of insulin receptor tyrosine kinase by metabolic intermediates. *Biochem. Soc. Trans.*, 17(6), 1110-1.

[160] Bennun, A.; Needle, M. A. and De Bari, V. A. (1985). Stimulation of the hexose monophosphate pathway in the human erythrocyte by Mn^{2+}: Evidence for a Mn^{2+}-dependent NADPH peroxidase activity. *Biochemical Medicine*, 33, 17-21.

[161] Vicario, P. P.; Saperstein, R. and Bennun, A. (1988). Role of divalent metals in the kinetic mechanism of insulin receptor tyrosine kinase. *Arch. Biochem. Biophys.*, 261(2), 336-45.

[162] Bennun A. (2012). Molecular Mechanisms Integrating Adenylyl Cyclase Responsiveness to Metabolic Control on Long-Term Emotional Memory and Associated Disorders. Nova Science Publishers, Inc. Long-Term Memory: Mechanisms, Types and Disorders (1-44). New York, USA.

[163] Ohanian, H.; Borhanian, K. and Bennun, A. (1978). The effect of manganese on the regulation of brain adenylate cyclase by magnesium and adenosine triphosphate. *Biochem. Soc. Trans.*, 6(6), 1179-82.

[164] Taqui-Khan, M. M. and Martell, A. E. (1962). Metal chelates of adenosine triphosphate. *J. Phys. Chem.*, 66, 10-15 (1962).

[165] Taqui-Khan, M. M.; Martell, A. E. (1966). Thermodynamic quantities associated with the interaction of adenosine triphosphate with metal ions. *J. Am. Chem. Soc.*, 88, 668-671.

[166] Bailey, C. H. and Kandel, E. R. (2008). Synaptic remodeling, synaptic growth and the storage of long-term memory in Aplysia. *Prog Brain Res.*, 169, 179-98.

[167] Reis, G. F.; Lee, M. B.; Huang, A. S. and Parfitt, K. D. (2005). Adenylate cyclase-mediated forms of neuronal plasticity in hippocampal area CA1 are reduced with aging. *J. Neurophysiol.*, 93(6), 3381-9.

[168] Veloso, D.; Guynn, R. W.; Oskarsson, M. and Veech, R. L. (1973). The concentrations of free and bound magnesium in rat tissues. Relative constancy of free Mg^{2+} concentrations. *J. Biol. Chem.*, 248 (13), 4811-4819.

[169] Rasmussen, H. (1970). Cell communication, calcium ion, and cyclic adenosine monophosphate. *Science*, 170 (3956), 404-412.

[170] Rasmussen, H. and Goodman D. B. (1975). Calcium and cAMP as interrelated intracellular messengers. *Ann. N.Y. Acad. Sci.*, 253, 789-796.

[171] León, D.; Castillo, C. A.; Ruiz, M. A.; Albasanz, J. L.and Martín., M. (2007). Metabotropic glutamate receptor/phospholipase C pathway is increased in rat brain at the end of pregnancy. *Neurochem Int.*, 50(5), 681-8.

[172] López-Téllez, J. F.; López-Aranda, M. F.; Navarro-Lobato, I.; Masmudi-Martín, M.; Martín Montañez, E.; Blanco Calvo E. and Khan, Z. U. (2010). Prefrontal Inositol Triphosphate Is Molecular Correlate of Working Memory in Nonhuman Primates. *The Journal of Neuroscience*, 30(8), 3067-3071.

[173] Strader, C. D.; Dixon, R. A.; Cheung, A. H.; Candelore, M. R.; Blake, A. D. and Sigal, I. S. (1987). Mutations that uncouple the beta-adrenergic receptor from Gs and increase agonist affinity. J. Biol. Chem., 262, 16439-43.

[174] Honore, T.; Lauridsen, J. and Krogsgaard-Larsen P. (1982). The binding of [3H]AMPA, a structural analogue of glutamic acid, to rat brain membranes. *Journal of Neurochemistry*, 38 (1), 173-178.

[175] Gray E. G. (1959). Axo-somatic and axo-dendritic synapses of the cerebral cortex: An electron microscopic study. *J. Anat.*, 93, 420-33.

[176] Johnston G. A. R. (1996). GABAA Receptor Pharmacology. *Pharmacology and Therapeutics*, 69 (3), 173-198.

[177] Chen, K.; Li, H. Z.; Ye, N.; Zhang, J. and Wang, J. J. (2005). Role of GABAB receptors in GABA and baclofen-induced inhibition of adult rat cerebellar interpositus nucleus neurons in vitro. *Brain Res Bull*, 67 (4), 310-8.

[178] Crispino, M.; Kaplan, B. B. and Martin R. (1997). Active polysomes are present in the large presynaptic endings of the synaptosomal fraction from squid brain. *J. Neurosci.*, 17, 7694-702.

[179] Crivellato, E.; Nico, B. and Ribatti, D. (2005). Ultrastructural evidence of piecemeal degranulation in larg dense-core vesicles of brain neurons. *Anat. Embryol. (Berl)*, 210, 25-34.

[180] De Camilli, P.; Slepnev, V. I.; Shupliakov, O. (2000) The estructure of synapses. In: Cowen W. M., Sudhof T. C., Stevens C. F., eds. *The structure of synapses.* Baltimore, M. D.: The Johns Hopkins University Press, 2000, 89-133.

[181] Deller, T.; Bas Orth, C.; Del Turco, D.; Vlachos, A.; Burbach, G. J.; Drakew, A.; Chabanis, S.; Korte, M.; Schwegler, H.; Haas, C. A. and Frotscher, M. (2007). A role for synaptopodin and the spine apparatus in hippocampal synaptic plasticity. *Ann. Anat.* 189 (1), 5-16.

[182] Fornai, F. (2007). Norepinephrine in neurological disorders. In: Ordway GA, Schwartz MA, Frazer A, editors. Brain Norepinephrine - Neurobiology and Therapeutics. Cambridge: University Press; pp. 436-71.

[183] Simpson, K. L. and Lin, R. C. S. (2007). Neuroanatomical and chemical organization of the locus Coeruleus. In: Ordway GA, Schwartz MA, Frazer A, editors. Brain Norepinephrine - Neurobiology and Therapeutics. Cambridge: University Press; pp. 9-52.

[184] Brodal, A. (1981). The reticular formation and some related nuclei. The nucleus locus coeruleus. In: Brodal A, editor. *Neurological anatomy in relation to clinical medicine.* New York: Oxford University Press; pp. 416-419.

[185] Nieuwenhuys, J. V.; Voogd, J. and vanHuijzen, C. The human central nervous system: a synopsis and atlas. *Berlin: Springer*, 1988.

[186] Fornai, F.; di Poggio, A. B.; Pellegrini, A.; Ruggieri, S. and Paparelli, A. (2007) Norepinephrine in Parkinson's disease: from disease progression to current therapeutics. *Curr. Med. Chem.*, 14, 2330-2334.

[187] Fornai, F.; Ruffoli, R.; Giorgi, F. S. and Paparelli, A. (2011). The role of locus coeruleus in the antiepileptic activity induced by vagus nerve stimulation. *Eur. J. Neurosci.*, 33, 2169-2178.

[188] Zhou, W.; Qian, Y., Kunjilwar, K., Pfaffinger, P. J., and Choe, S. (2004) Structural insights into the functional interaction of KChIP1 with Shal-type K(+) channels. *Neuron*. 41 (4), 573-586.

[189] Norris, J. G. and Benveniste, E. N. (1993). Interleukin-6 production by astrocytes: induction by the neurotransmitter norepinephrine. *J. Neuroimmunol.* 45, 137-145.

[190] Kingham, P. J. and Pocock, J. M. (2000). Microglial apoptosis induced by chromograinin A is mediated by mitochondrial depolarisation and the permeability transition but not by cytochrom C release. *J. Neurochem.*, 74, 1452-1462.

[191] Engelhardt, B. and Sorokin, L. (2009). The blood-brain and the blood-cerebrospinal fluid barriers: function and dysfunction. *Semin Immunopathol.*, 31(4), 497-511

[192] Buxbaum, J. D.; Choi, E. K.; Luo, Y.; Lilliehook, C.; Crowley, A. C.; Merriam, D. E. and Wasco, W. (1998). Calsenilin: a calcium-binding protein that interacts with the presenilins and regulates the levels of a presenilin fragment. *Nat. Med.*, 4 (10), 1177-1181.

[193] An, W. F., Bowlby, M. R., Betty, M., Cao, J., Ling, H. P., Mendoza, G., Hinson, J. W., Mattsson, K. I., Strassle, B. W., Trimmer, J. S., and Rhodes, K. J. (2000). Modulation of A-type potassium channels by a family of calcium sensors. *Nature*, 403 (6769), 553-556.

[194] Carrion, A. M.; Mellstrom, B. and Naranjo, J. R. (1998). Protein kinase A-dependent derepression of the human prodynorphin gene via differential binding to an intragenic silencer element. *Mol. Cell. Biol.*, 18 (12), 6921-6929.

[195] Ledo, F.; Link, W. A.; Carrion, A. M.; Echeverria, V.; Mellstrom, B. and Naranjo, J. R. (2000). The DREAM-DRE interaction: key nucleotides and dominant negative mutants. *Biochim. Biophys. Acta*, 1498 (2-3), 162-168.

[196] Osawa, M.; Dace, A.; Tong, K. I.; Valiveti, A.; Ikura, M. and Ames, J. B. (2005). Mg^{2+} and Ca^{2+} Differentially Regulate DNA Binding and Dimerization of DREAM. *J Biol Chem*. 280 (18), 18008-14.

[197] Bekar, L. K.; Wei, H. S. and Nedergaard, M. (2012). The locus coeruleus-norepinephrine network optimizes coupling of cerebral blood volume with oxygen demand. *J. Cereb Blood Flow Metab.*, 32(12), 2135-45.

[198] Wang, F.; Smith, N.A.; Xu, Q.; Fujita, T.; Baba, A.; Matsuda, T.; Takano, T.; Bekar, L. and Nedergaard, M. (2012). Astrocytes modulate

neural network activity by Ca^{2+}-dependent uptake of extracellular K^+. *Sci. Signal.*, 5(218), ra26.

[199] Goldin, A. L.; Barchi, R. L.; Caldwell, J. H.; Hofmann, F.; Howe, J. R.; Hunter, J. C.; Kallen, R. G.; Mandel, G.; Meisler, M. H.; Netter, Y. B.; Noda, M.; Tamkun, M. M.; Waxman, S. G.; Wood. J. N. and Catterall, W. A. (2000). Nomenclature of voltage-gated sodium channels. *Neuron*, 28(2), 365-8.

[200] Barchi, R. L. (1998). Ion channel mutations affecting muscle and brain. *Curr Opin Neurol.*, 11(5), 461-8.

[201] Sun, W.; Barchi, R. L. and Cohen, S. A. (1995). Probing sodium channel cytoplasmic domain structure. Evidence for the interaction of the rSkM1 amino and carboxyl termini. *J. Biol. Chem.*, 270 (38), 22271-6.

[202] Kleopa, K. A. and Barchi R. L. (2002). Genetic disorders of neuromuscular ion channels. *Muscle Nerve.*, 26 (3), 299-325.

[203] Kraner, S. D.; Rich, M. M.; Sholl, M. A.; Zhou, H.; Zorc, C. S.; Kallen, R. G. and Barchi, R. L. (1999). Interaction between the skeletal muscle type 1 Na+ channel promoter E-box and an upstream repressor element. Release of repression by myogenin. *J. Biol Chem.*, 274 (12), 8129-36.

[204] Barchi, R. L.; Rich, M. M. and Kraner, S. D. (1999). Altered gene expression in steroid-treated denervated muscle. *Neurobiol Dis.*, 6 (6), 515-22.

[205] Neuwelt, E. A.; Bauer, B.; Fahlke, C.; Fricker, G.; Iadecola, C.; Janigro, D.; Leybaert, L.; Molnár, Z.; O'Donnell, M. E.; Povlishock, J. T.; Saunders, N. R.; Sharp, F.; Stanimirovic, D.; Watts, R. J. and Drewes, L. R. (2011). Engaging neuroscience to advance translational research in brain barrier biology. *Nat. Rev. Neuroscience*, 12 (3), 169-182.

[206] Bourtchuladze, R.; Frenguelli, B.; Blendy, J.; Cioffi, D.; Schutz, G. and Silva, A. J. (1994). Deficient long-term memory in mice with a targeted mutation of the cAMP-responsive element-binding protein. *Cell*, 79 (1), 59-68.

[207] Mørk, A. and Geisler, A. (1987). Effects of lithium on calmodulin-stimulated adenylate cyclase activity in cortical membranes from rat brain. *Pharmacology and Toxicology*, 60 (1), 17-23.

[208] Jung, J. S.; Bhat, R. V.; Preston, G. M.; Guggino, W. B.; Baraban, J. M. and Agre, P. (1994). Molecular characterization of an aquaporin cDNA from brain: candidate osmoreceptor and regulator of water balance. *Proc Natl Acad Sci USA*, 91 (26), 13052-6.

[209] Musa-Aziz, R.; Chen, L. M.; Pelletier, M. F. and Boron, W. F. (2009). Relative CO2/NH3 selectivities of AQP1, AQP4, AQP5, AmtB, and RhAG. *Proc. Natl. Acad. Sci. USA*., 106 (13), 5406-11.

[210] Benfenati, V. and Ferroni, S. (2010). Water transport between CNS compartments: functional and molecular interactions between aquaporins and ion channels. *Neuroscience*, 168, 926-940.

[211] Cerdan, S. and Seelig, J. (1990). NMR studies of metabolism. *Annu. Rev. Biophys. Biophys. Chem*., 19, 43-67.

[212] Kunnecke, B. (1995). Application of ^{13}C NMR spectroscopy to metabolic studies on animals, in *Carbon-13 NMR Spectroscopy of Biological Systems*, ed. N. Beckman (New York: Academic Press, Inc.), 159–267.

[213] Morris, P., and Bachelard, H. (2003). Reflections on the application of C-13-MRS to research on brain metabolism. *NMR Biomed*. 16, 303-312.

[214] Rodrigues, T. B., and Cerdán, S. (2007). The cerebral tricarboxylic acid cycles. In Handbook of Neurochemistry and Molecular Neurobiology, eds A. Lajtha, G. Gibson, and G. Dienel (New York: Springer), 63-91.

[215] Rodrigues, T. B.; Valette, J. and Anne-Karine Bouzier-Sore. (2013). ^{13}C NMR spectroscopy applications to brain energy metabolism. *Front. Neuroenergetics*, 5, 9.

[216] Cerdan, S.; Sierra, A.; Fonseca, L. L.; Ballesteros, P. and Rodrigues, T. B. (2009). The turnover of the H3 deuterons from (2-13C) glutamate and (2-13C) glutamine reveals subcellular trafficking in the brain of partially deuterated rats. *J. Neurochem*. 109(Suppl. 1), 63-72.

[217] Friebolin, H. (1991). Basic One- and Two-dimensional NMR Spectroscopy. Verlagsgesellschaft ; New York, NY, USA : VCH Publishers.

[218] Chowdhury, G. M. I.; Patel, A. B.; Mason, G. F.; Rothman, D. L. and Behar K. L. (2007) Glutamatergic and GABAergic neurotransmitter cycling and energy metabolism in rat cerebral cortex during postnatal development. *J Cereb Blood Flow Metab*, 27(12), 1895-1907.

[219] Brekke, E. M.; Walls, A. B.; Schousboe, A.; Waagepetersen, H. S. and Sonnewald, U. (2012) Quantitative importance of the pentose phosphate pathway determined by incorporation of 13C from [2-13C]- and [3-13C]glucose into TCA cycle intermediates and neurotransmitter amino acids in functionally intact neurons. *J Cereb Blood Flow Metab*, 32 (9), 1788-1799.

[220] Norenberg, M. D. and Martinez-Hernandez, A. (1979). Fine structural localization of glutamine synthetase in astrocytes of rat brain. *Brain Res.* 161, 303-310.

[221] Ozkan, E. D., and Ueda, T. (1998). Glutamate transport and storage in synaptic vesicles. *Jpn. J. Pharmacol.* 77, 1-10.

[222] Amaral, A. I., Teixeira, A. P., Sonnewald, U., and Alves, P. M. (2011). Estimation of intracellular fluxes in cerebellar neurons after hypoglycemia: importance of the pyruvate recycling pathway and glutamine oxidation. *J. Neurosci. Res.* 89, 700-710.

[223] Birnbaum, S. M.; Levintow, L.; Kingsley, R. B. and Greenstein, J. P. (1952). Specificity of amino acid acylases. *J. Biol. Chem.* 194: 455-470.

[224] Miller, D. S. (2010). Regulation of P-glycoprotein and other ABC drug transporters at the blood-brain barrier. *Trends Pharmacol. Sci.* 31, 246-254.

[225] Abbott, N. J.; Rönnbäck, L. and Hansson, E. (2006). Astrocyte-endothelial interactions at the blood–brain barrier. *Nature Rev. Neurosci.*, 7 (1), 41-53.

[226] Iadecola, C. and Nedergaard, M. (2007). Glial regulation of the cerebral microvasculature. *Nature Neurosci.*, 10, 1369-1376.

[227] Kimelberg, H. K. and Nedergaard, M. Functions of astrocytes and their potential as therapeutic targets. *Neurotherapeutics*, 7, 338-353 (2010).

[228] Simard, M.; Arcuino, G.; Takano, T.; Liu, Q. S. and Nedergaard, M. (2003). Signalling at the gliovascular interface. *J. Neurosci.*, 23, 9254-9262.

[229] Davidson, A. L.; Dassa, E.; Orelle, C. and Chen, J. (2008). Structure, function, and evolution of bacterial ATP-binding cassette systems. *Microbiol. Mol. Biol. Rev.*, 72 (2), 317-64.

[230] Goffeau, A.; de Hertogh, B. and Baret, P. V. (2004). ABC Transporters. *In: Encyclopedia of Biological Chemistry*, 1, 1-5.

[231] Jones, P. M. and George, A. M. (2004). The ABC transporter structure and mechanism: perspectives on recent research. *Cell Mol Life Sci.*, 61 (6), 682-99.

[232] Ponte-Sucre, A. (2009). ABC transporters in microorganisms. Caister Academic Press. ISBN 978-1-904455-49-3.

[233] Perea, G.; Navarrete, M. and Araque, A. (2009). Tripartite synapses: astrocytes process and control synaptic information. *Trends. Neurosci.* 32, 421-431.

[234] Gadsby, D. C.; Bezanilla, F.; Rakowski, R. F.; De Weer, P. and Holmgren, M. (2012). The dynamic relationships between the three

events that release individual Na^+ ions from the Na^+/K^+-ATPase. *Nature Cell Biology*, 14, 416-423.

[235] Prigogine, I.; Lefever, R.; Goldbeter, A. and Herschkowitz-Kaufman, M. (1969). Symmetry breaking instabilities in biological systems. *Nature*, 223 (5209), 913-6.

[236] Thompson, S. W.; Dray, A. and Urban, L. (1994). Injury-induced plasticity of spinal reflex activity: NK1 neurokinin receptor activation and enhanced A- and C-fiber mediated responses in the rat spinal cord in vitro. *J Neurosci.*, 14(6), 3672-87.

[237] Guinzel, D. and Schlue, W. R. (1996). Sodium-magnesium antiport in Retzius neurones of the leech Hirudo medicinalis. *Journal of Physiology*, 491 (3), 595-608.

[238] Jardetzky, O. (1966). Simple allosteric model for membrane pumps. *Nature*, 211, 969-970.

[239] Clausen, J. D.; McIntosh, D. B.; Woolley, D. G. and Andersen, J. P. (2011). Modulatory ATP binding affinity in intermediate states of E2P dephosphorylation of sarcoplasmic reticulum Ca^{2+}-ATPase. *J. Biol. Chem.*, 286, 11792-11802.

[240] Jensen, A. M.; Sorensen, T. L.; Olesen, C.; Moller, J. V. and Nissen, P. (2006). Modulatory and catalytic modes of ATP binding by the calcium pump. *EMBOJ.*, 25, 2305-2314.

[241] Bublitz, M.; Morth, J. P. and Nissen, P. (2011). P-type ATPases at a glance. *J. Cell Sci.* 124(Pt 15), 2515-9.

[242] Albers, R. W. (1967). Biochemical aspects of active transport. *Annu. Rev. Biochem.*, 36, 727-756.

[243] Olesen, C.; Picard, M.; Winther, A. M.; Gyrup, C.; Morth, J. P.; Oxvig, C.; Moller, J. V. and Nissen, P. (2007). The structural basis of calcium transport by the calcium pump. *Nature*, 450, 1036-1042.

[244] Raben, M. S. and Hollemberg, C. H. (1959). Effect of growth hormone on plasma fatty acids. *J. Clin. Invest.*, 38(3), 484-8.

[245] Lefkowitz, R. J. (2007). Seven transmembrane receptors: something old, something New. *Acta Physiol. (Oxf.)*, 190(1), 9-19.

[246] William, R. (2010). Robert Lefkowitz: godfather of G protein-coupled receptors. *Circ. Res.*, 106 (5), 812-4.

[247] Mazzocchi, G.; Malendowicz, L. K.; Gottardo, L.; Aragona, F. and Nussdorfer, G. G. (2001). Orexin A stimulates cortisol secretion from human adrenocortical cells through activation of the adenylate cyclase-dependent signaling cascade. *J. Clin. Endocrinol. Metab.*, 86 (2), 778-82.

[248] Sutherland, E. W.; Rall, T. W. and Menon, T. (1962). Adenylyl cyclase. I. Distribution, preparation, and properties. *J. Biol. Chem.*, 237, 1220-7.

[249] Pupim, L. B.; Flakoll, P. J.; Yu, C. and Ikizler, T. A. (2005). Recombinant human growth hormone improves muscle amino acid uptake and whole-body protein metabolism in chronic hemodialysis patients. *Am. J. Clin. Nutr.*, 82 (6), 1235-43.

[250] Grimm, S.; Pestke, K.; Feeser, M.; Aust, S.; Weigand, A.; Wang, J.; Wingenfeld, K.; Pruessner, J.C.; Böker, H. and Bajbouj, M. (2014). Early life stress modulates oxytocin effects on limbic system during acute psychosocial stress. *Soc Cogn Affect Neurosci.* Jan 28. [Epub ahead of print]

[251] Taelman, V. F.; Dobrowolski, R.; Plouhinec, J. L.; Fuentealba, L. C.; Vorwald, P. P.; Gumper, I.; Sabatini, D. D. and De Robertis, E. M. (2010). Wnt signaling requires sequestration of glycogen synthase kinase 3 inside multivesicular endosomes. *Cell.*, 143(7), 1136-48.

[252] Zeigerer, A.; Lampson, M. A.; Karylowski, O.; Sabatini, D. D.; Adesnik, M.; Ren, M. and McGraw, T. E. (2002). GLUT4 retention in adipocytes requires two intracellular insulin-regulated transport steps. *Mol. Biol. Cell.*, 13 (7), 2421-35.

[253] Barbacid, M. (1994). The Trk family of neurotrophin receptors. *J. Neurobiol.*, 25(11), 1386-403.

[254] Rapp, G.; Klaudiny, J.; Hagendorff, G.; Luck, M. R. and Scheit, K. H. (1989). Complete sequence of the coding region of human elongation factor 2 (EF-2) by enzymatic amplification of cDNA from human ovarian granulosa cells. *Biol. Chem. Hoppe Seyler.*, 370 (10), 1071-5.

[255] Vicario, P. P.; Saperstein, R. and Bennun, A. (1988). Regulation of insulin receptor tyrosine kinase by divalent metal cations, metal-ATP substrate and free ATP4-. Biochem. Soc. Trans., 16, 40-42.

[256] Choe, E.S. and Wang, J. Q. (2002). Regulation of transcription factor phosphorylation by metabotropic glutamate receptor-associated signaling pathways in rat striatal neurons. *Neuroscience*, 114(3),557-65.

[257] Ahmed, T. and Frey J. U. (2005). Plasticity-specific phosphorylation of CaMKII, MAP-kinases and CREB during late-LTP in rat hippocampal slices in vitro. *Neuropharmacology*. 49(4), 477-92.

[258] Chawla, S. and Bading, H. (2001). CREB/CBP and SRE-interacting transcriptional regulators are fast on-off switches: duration of calcium transients specifies the magnitude of transcriptional responses. *J. Neurochem., 79(4), 849-58.*

[259] Rana, S.; Blowers, E. C. and Natarajan, A. (2014 Aug 28). Small molecule adenosine 5′-monophosphate activated protein kinase (AMPK) modulators and human diseases. *J. Med. Chem.* [Epub ahead of print]

[260] Impey, S.; McCorkle, S. R.; Cha-Molstad, H.; Dwyer, J.M.; Yochum, G. S.; Boss, J. M.; McWeeney, S.; Dunn, J. J.; Mandel, G. and Goodman, R. H. (2004). Defining the CREB regulon: a genome-wide analysis of transcription factor regulatory regions. *Cell.*, 119(7), 1041-54.

[261] Alberini, C. M. (2009) Transcription Factors in Long-Term Memory and Synaptic Plasticity. Physiological Reviews, 89(1), 121-145.

[262] Carlezon, W. A. Jr.; Duman, R. S. and Nestler, E. J. (2005). The many faces of CREB. *Trends Neurosci.* 28(8), 436-45.

[263] Finkbeiner, S. (2001). New roles for introns: sites of combinatorial regulation of Ca^{2+}- and cyclic AMP-dependent gene transcription. *Sci. STKE.* 2001(94), pe1.

[264] Kornhauser, J. M.; Cowan, C. W.; Shaywitz, A. J.; Dolmetsch, R. E.; Griffith, E. C.; Hu, L. S.; Haddad, C.; Xia, Z. and Greenberg, M. E. (2002) CREB transcriptional activity in neurons is regulated by multiple, calcium-specific phosphorylation events. *Neuron. 34(2), 221-33.*

[265] Husi, H.; Ward, M. A.; Choudhary, J.S.; Blackstock, W. P. and Grant, S. G. (2000). Proteomic analysis of NMDA receptor-adhesion protein signaling complexes. *Nat. Neurosci.*, 3(7), 661-9.

[266] Chan, C. B.; Liu, X.; He, K.; Qi, Q.; Jung, D. Y.; Kim, J. K. and Ye, K. (2011). The association of phosphoinositide 3-kinase enhancer A with hepatic insulin receptor enhances its kinase activity. *EMBO Rep.*, 12 (8), 847-54.

[267] Lindsley, J. E. and Rutter, J. (2004). Nutrient sensing and metabolic decisions. *Comp Biochem Physiol B Biochem Mol Biol.*, 139 (4), 543-59.

[268] Boudeau J.; Scott, J. W.; Resta, N.; Deak, M.; Kieloch, A.; Komander, D.; Hardie, D. G.; Prescott, A. R.; van Aalten, D. M. and Alessi, D. R. (2004). Analysis of the LKB1-STRAD-MO25 complex. *J. Cell Sci.*, 117(Pt 26), 6365-75.

[269] Hardie, D. G.; Scott, J. W.; Pan, D. A. and Hudson, E. R. (2003). Management of cellular energy by the AMP-activated protein kinase system. *FEBS Lett.* 546(1), 113-20.

[270] Michiels, C. (2004). Physiological and pathological responses to hypoxia. *Am. J. Pathol.*, 164(6), 1875-82.

[271] Forcet, C.; Etienne-Manneville, S.; Gaude, H.; Fournier, L.; Debilly, S.; Salmi, M.; Baas, A.; Olschwang, S.; Clevers, H. and Billaud, M. (2005). Functional analysis of Peutz-Jeghers mutations reveals that the LKB1 C-

terminal region exerts a crucial role in regulating both the AMPK pathway and the cell polarity. *Hum Mol Genet.*, 14(10), 1283-92.

[272] Atkinson, D. E. (1968). The energy charge of the adenylate pool as a regulatory parameter. Interaction with feedback modifiers. *Biochemistry*, 7 (11), 4030-4034.

[273] Tinbergen N. (1974). Ethology and Stress Diseases. *Science.*, 185 (4145), 20-27.

[274] Bennun A. and Avron M. (1964). Light-dependent and light-triggered adenosine triphosphatases in chloroplasts. *Biochim Biophys Acta*, 79, 646-8.

[275] Bennun A. and Avron M. (1965). The relation of the light-dependent and light-triggered adenosine triphosphatases to photophosphorylation. *Biochim. Biophys. Acta*, 109 (1), 117-27.

[276] Benarroch, E. E. (2009). The locus ceruleus norepinephrine system: functional organization and potential clinical significance. *Neurology.*, 73(20), 1699-704.

[277] Adolphs, R.; Cahill, L.; Schul, R. and Babinsky, R. (1997). Impaired declarative memory for emotional material following bilateral amygdala damage in humans. *Learning & Memory*, 4, 291-300.

[278] Cahill, L.; Babinsky, R.; Markowitsch, H. J. and McGaugh, J. L. (1995). The amygdala and emotional memory. *Nature*, 377 (6547), 295-296.

[279] Collingridge, G. L.; Peineau, S.; Howland, J. G. and Wang, Y. T. (2010). Long-term depression in the CNS. *Nat Rev Neurosci.*, 11(7), 459-73.

[280] Salmans, S. (1997). Depression: Questions You Have – Answers You Need. People's Medical Society. ISBN 978-1-882606-14-6.

[281] Pertwee, R. G. (April 2006). The pharmacology of cannabinoid receptors and their ligands: an overview. *Int J Obes (Lond)*, 30 (1), S13-8.

[282] Bluett, R. J.; Gamble-George, J. C.; Hermanson, D. J.; Hartley, N. D.; Marnett, L. J. and Patel, S. (2014). Central anandamide deficiency predicts stress-induced anxiety: behavioral reversal through endocannabinoid augmentation. *Translational Psychiatry*, 4, e408.

[283] Moreira, F. A.; Grieb, M. and Lutz, B. (2009). Central side-effects of therapies based on CB1 cannabinoid receptor agonists and antagonists: focus on anxiety and depression. *Best Pract Res Clin Endocrinol Metab*, 23 (1), 133-144.

[284] Bennun, A. and Blum, J. J. (1966). Properties of the induced acid phosphatase and of the constitutive acid phosphatase of Euglena. *Biochim. Biophys. Acta*, 128(1), 106-23.

[285] Carlson, N. (2013). *Physiology of Behavior*. Pearson. pp. 602-606.

[286] Porterfield, V. M.; Gabella, K. M.; Simmons, M. A. and Johnson, J. D. (2012). Repeated stressor exposure regionally enhances beta-adrenergic receptor-mediated brain IL-1β production. *Brain Behav. Immun.*, 26(8), 1249-55.

[287] Talbot, K.; Wang, H. Y.; Kazi, H.; Han, L. Y.; Bakshi, K. P.; Stucky, A.; Fuino, R. L.; Kawaguchi, K. R.; Samoyedny, A. J.; Wilson, R. S.; Arvanitakis, Z.; Schneider, J. A.; Wolf, B. A.; Bennett, D. A.; Trojanowski, J. Q. and Arnold, S. E. (2012). Demonstrated brain insulin resistance in Alzheimer's disease patients is associated with IGF-1 resistance, IRS-1 dysregulation, and cognitive decline. *J Clin Invest.*, 122(4), 1316-1338.

[288] Van Assema, D. M.; Goos, J. D.; van der Flier, W. M.; Lubberink, M.; Boellaard, R.; Windhorst, A. D.; Scheltens, P.; Lammertsma, A. A. and van Berckel, B. N. (2012). No evidence for additional blood-brain barrier P-glycoprotein dysfunction in Alzheimer's disease patients with microbleeds. *J Cereb Blood Flow Metab.*, 32(8), 1468-71.

[289] Sahin, K.; Tuzcu, M.; Orhan, C.; Ali, S.; Sahin, N.; Gencoglu, H.; Ozkan, Y.; Hayirli, A.; Gozel, N. and Komorowski, J. R. (2013). Chromium modulates expressions of neuronal plasticity markers and glial fibrillary acidic proteins in hypoglycemia-induced brain injury. *Life Sci.*, 93 (25-26), 1039-48.

[290] Walter, P. and Ron, D. (2011). The unfolded protein response: from stress pathway to homeostatic regulation. *Science*, 334 (6059), 1081-6.

[291] Bennun, A. (2012). The dynamics of H-bonds of the hydration shells of ions, ATPase and NE-activated adenylyl cyclase on the coupling of energy and signal transduction. *arXiv:1208.5673v1 [q-bio.OT].*

[292] Mackliff, J.; Sanchez O. and Bennun, A. (2013). Bilateral electrocoagulation of adrenal medulla, a surgical neurotechnology applied to the dysfunctional hypothalamic-pituitary-adrenal axis. *Accelerating Translational Neurotechnology Fourth Annual Aspen Brain Forum. Aspen Meadows Resort.*

[293] Bennun, A. (2013) A nano-insert silicate Sr-89 applied to decrease the hyper-functionality of micro-sized tissue. *Accelerating Translational Neurotechnology. Sept 18-20, 2013 Aspen, CO.*

[294] Bennun A. (2012). Recovery of radioisotopes from nuclear waste for radio-scintillator-luminescence energy applications. *International Journal of Energy, Environment, and Economics, Nova Science Publishers, Inc.*, 20 (5), 509-515.

[295] Harris, L. W.; Guest, P. C.; Wayland, M. T.; Umrania, Y.; Krishnamurthy, D.; Rahmoune, H. and Bahn, S. (2013). Schizophrenia: metabolic aspects of aetiology, diagnosis and future treatment strategies. *Psychoneuroendocrinology*, 38 (6), 752-66.

[296] Ohki, K.; Chung, S.; Kara, P.; Hübener, M.; Bonhoeffer, T. and Reid, R.C. (2006). Highly ordered arrangement of single neurons in orientation pinwheels. *Nature*, 442 (7105), 925-8.

[297] Ballabh, P.; Braun, A. and Nedergaard, M. (2004). The blood-brain barrier: an overview: structure, regulation, and clinical implications. *Neurobiol. Dis.*, 16 (1), 1-13.

[298] Li, F. and Tsien, J. Z. (2009). Memory and the NMDA receptors. *N. Engl. J. Med.*, 361 (3), 302-3.

[299] Slutsky, I.; Sadeghpour, S.; Li, B. and Liu, G. (2004). Enhancement of synaptic plasticity through chronically reduced Ca^{2+} flux during uncorrelated activity. *Neuron.*, 44 (5), 835-49.

[300] Wei, Y. Z.; Kumbharkhane, A. C.; Sadeghi, M.; Sage, J. T.; Tian, W. D.; Champion, P. M.; Sridhar, S. and McDonald, M. J. (1994). Protein hydration investigations with high-frequency dielectric spectroscopy. *J. Phys. Chem.*, 98 (26), 6644-51.

[301] Barco, A.; Balley, C. H. and Kandel, E. R. (2006). Common molecular mechanisms in explicit and implicit memory. *J. Neurochem.*, 97 (6), 1520-33.

[302] Bean, B. P. (2007). The action potential in mammalian central neurons. *Nat. Rev. Neurosci.*, 8 (6), 451-65.

[303] Bennun, A. (2012). The dynamics of H-bonds of the hydration shells of ions, ATPase and NE-activated adenylyl cyclase on the coupling of energy and signal transduction. *Nova Science Publishers, Inc. International Journal of Medical and Biological Frontiers*, 18 (11), 767-782).

[304] Bennun, A. (1974). A model mechanism for coupled phosphorylation. *Proc. 3rd Int. Cong. Photosynthesis, Rehovoth (M. Avron, ed.)*, Vol. 2, 1107-1120, *Elsevier Sci. Pub. Co., Amsterdam.*

[305] Bennun A. and Bennun, N. Hypothesis for a mechanism of energy transduction. Sigmoidal kinetics of chloroplast's heat-activated ATPase. Proceedings 2nd International Congress on Photosynthesis Research (G. Fortí, M. Avron and A. Melardri, eds.), 2, 1115-1124, Dr. W. Junk N.V. Pub., The Hague (1972).

[306] Lomize A. L.; Pogozheva I. D. and Mosberg, H. I. (2011). Anisotropic solvent model of the lipid bilayer. 2. Energetics of insertion of small

molecules, peptides and proteins in membranes. *J. Chem. Inf. Model.*, 51 (4), 930-46.

[307] Gibb, C. L. and Gibb, B. C. (2011) Anion binding to hydrophobic concavity is central to the salting-in effects of Hofmeister chaotropes. *J. Am. Chem. Soc.*, 133 (19), 7344-7.

[308] Schmidt, D. A.; Birer, O.; Funkner, S.; Born B. P.; Gnanasekaran, R.; Schwaab, G. W.; Leitner, D. M. and Havenith, M. (2009). Rattling in the cage: ions as probes of sub-picosecond water network dynamics. *J. Am. Chem. Soc.*, 131 (51), 18512-7.

[309] Ding, F.; O'Donnell, J.; Thrane, A. S.; Zeppenfeld, D.; Kang, H.; Xie, L.; Wang, F. and Nedergaard, M. (2013). α1-Adrenergic receptors mediate coordinated Ca(2+) signaling of cortical astrocytes in awake, behaving mice. *Cell Calcium.*, 54 (6), 387-94.

[310] Barchi, R. L. (1993). Ion channels and disorders of excitation in skeletal muscle. *Curr. Opin. Neurol. Neurosurg.*, 6 (1), 40-7.

[311] Schotland, D. L.; Fieles, W. and Barchi, R. L. (1991). Expression of sodium channel subtypes during development in rat skeletal muscle. *Muscle Nerve*, 14(2), 142-51.

[312] Purves, D.; Augustine, G. J.; Fitzpatrick, D.; Hall, W. C.; LaMantia, A. S.; McNamara, J. O. and White, L. E. (2008). *Neuroscience*. 4th ed. Sinauer Associates. pp. 170-6.

[313] Moll, J. R.; Acharya, A.; Gal, J.; Mir A. A. and Vinson C. (2012). Magnesium is required for specific DNA binding of the CREB B-ZIP domain. *Nucleic Acids Res.*, 30(5), 1240-6.

[314] Lawson, C. L.; Swigon, D.; Murakami, K. S.; Darst, S. A.; Berman, H. M. and Ebright, R. H. (2004). Catabolite activator protein: DNA binding and transcription activation. *Curr Opin Struct Biol.*, 14(1), 10-20.

[315] Amsten, A. F. T. (2011) Prefrontal cortical network connections: key site of vulnerability in stress and schizophrenia. *International Journal of Developmental Neuroscience*, 29(3), 215-223.

[316] Goelet, P.; Castellucci, V. F.; Schacher, S. and Kandel, E. R. (1986). The long and the short of long-term memory a molecular framework. *Nature*, 322(6078), 419-22.

[317] Eccles, J. C. (1983). Calcium in long-term potentiation as a model for memory. *Neuroscience*, 10(4), 1071-81.

[318] Silva, A. J.; Kogan, J. H.; Frankland, P. W. and Kida, S. (1998). CREB and memory. *Annu. Rev. Neurosci.*, 21, 127-48.

[319] Davis, H. P. and Squire, L. R. (1984). Protein synthesis and memory: a review. *Psychol. Bull.*, 96 (3), 518-59.

[320] Mayr, B. and Montminy, M. (2001). Transcriptional regulation by the phosphorylation dependent factor CREB. *Nat. Rev. Mol. Cell Biol.*, 2 (8), 599-609.

[321] Kandel, E. R. (2012). The molecular biology of memory: cAMP, PKA, CRE, CREB-1, CREB-2, and CPEB. *Mol. Brain*, 5, 14.

[322] Kandel, E. R. (2001). The molecular biology of memory storage: a dialogue between genes and synapses. *Science*, 294 (5544), 1030-8.

[323] Greenspan, R. J.; Tononi, G.; Cirelli, C. and Shaw, P. J. (2001). Sleep and the fruit fly. *Trends Neurosci.*, 24, 142-145.

[324] Malow, B. A. (2004). Sleep deprivation and epilepsy. *Epilepsy Curr.*, 4(5), 193-5.

[325] Stickgold, R. (2006). Neuroscience: a memory boost while you sleep. *Nature*, 444(7119), 559-60.

[326] Tonomi, G. (2014). Sleep and the price of plasticity: from synaptic and cellular homeostasis to memory consolidation and integration. *Neuron* (Impact Factor: 15.77), 81(1), 12-34.

[327] Kress, B. T.; Iliff, J. J.; Xia, M.; Wang, M.; Wei, H. S.; Zeppenfeld, D.; Xie, L.; Kang, H.; Xu, Q.; Liew, J. A.; Plog, B. A.; Ding, F.; Deane, R. and Nedergaard, M. (2014). Impairment of paravascular clearance pathways in the aging brain. *Ann. Neurol.*, 76(6), 845-61.

[328] Iliff, J. J.; Wang, M.; Liao, Y.; Plogg, B. A.; Peng, W.; Gundersen, G. A.; Benveniste, H.; Vates, G. E.; Deane, R.; Goldman, S. A.; Nagelhus, E. A. and Nedergaard M. (2012). A paravascular pathway facilitates CSF flow through the brain parenchyma and the clearance of interstitial solutes, including amyloid β. *Sci Transl Med.*, 4(147), 147ra111.

[329] Mendelsohn, A. R. and Larrick, J. W. (2013) Sleep facilitates clearance of metabolites from the brain: glymphatic function in aging and neurodegenerative diseases. *Rejuvenation Res.*, 16(6), 518-23.

[330] Constantinople, C. M. and Bruno, R. M. (2011). Effects and mechanisms of wakefulness on local cortical networks. *Neuron.*, 69(6), 1061-8.

[331] O'Donnell, J.; Zeppenfeld, D.; McConnell, E.; Pena, S. and Nedergaard, M. (2012). Norepinephrine: a neuromodulator that boosts the function of multiple cell types to optimize CNS performance. *Neurochem Res.*, 37(11), 2496-512.

[332] McBain, C. J.; Traynelis, S. F. and Dingledine, R. (1990). Regional variation of extracellular space in the hippocampus. *Science.*, 249(4969), 674-7.

[333] Wang, F.; Smith, N. A.; Xu, Q.; Goldman, S.; Peng, W.; Huang, J. H.; Takano, T. and Nedergaard, M. (2013). Photolysis of caged Ca^{2+} but not

receptor-mediated Ca^{2+} signaling triggers astrocytic glutamate release. *J. Neurosci.*, 33(44), 17404-12.

Author Contact Information:

Dr. Alfred Bennun, Full Professor of Biochemistry
Emeritus - Rutgers University
University Connection Card/ID Office Information
647 East 14th Street, Apartment 2G Expiration: Indefinite
New York, NY 10009-3158

Consultant, CONICET, AR
Adress: Godoy Cruz 3046
Tower 2 – 8o Polo – Buenos Aires city 1425
Tel & Fax: (011) 4775-7286 alfr9@hotmail.com

Index

#

A

B

C

D

E

F

G

H

I

J

K

L

M

N

O

P

R

S

T

U

V

W

X

Y

Z

α

β

γ